Comparative Dental Morphology

Frontiers of Oral Biology

Vol. 13

Selected papers of the 14th International Symposium on Dental Morphology, August 27–30, 2008, Greifswald, Germany

Comparative Dental Morphology

Volume Editors

T. Koppe Greifswald
G. Meyer Greifswald
K.W. Alt Mainz

Co-Editors

A. Brook Liverpool
M.C. Dean London
I. Kjær Copenhagen
J.R. Lukacs Eugene, Oreg.
B.H. Smith Ann Arbor, Mich.
M.F. Teaford Baltimore, Md.

61 figures, 10 in color, and 20 tables, 2009

KARGER

Basel · Freiburg · Paris · London · New York · Bangalore · Bangkok · Shanghai · Singapore · Tokyo · Sydney

Frontiers of Oral Biology

Prof. Dr. Thomas Koppe
Institut für Anatomie und Zellbiologie
Ernst-Moritz-Arndt-Universität
Greifswald
Friedrich-Loeffler-Str. 23c
DE-17487 Greifswald (Germany)

Prof. Dr. Georg Meyer
Poliklinik für Zahnerhaltung,
Parodontologie und Endodontologie
Ernst-Moritz-Arndt-Universität
Greifswald
Rotgerberstraße 8
DE-17475 Greifswald (Germany)

Prof. Dr. Kurt W. Alt
Institut für Anthropologie
Johannes-Gutenberg-Universität
Mainz
Colonel Kleinmann Weg 2
DE-55099 Mainz (Germany)

Library of Congress Cataloging-in-Publication Data

International Symposium on Dental Morphology (14th : 2008 : Greifswald, Germany)
Comparative dental morphology / volume editors, T. Koppe, G. Meyer, K.W. Alt ; co-editors, A. Brook ... [et al.].
p. ; cm. -- (Frontiers of oral biology, ISSN 1420-2433 ; v. 13)
Includes bibliographical references and indexes.
ISBN 978-3-8055-9229-1 (hard cover : alk. paper)
1. Teeth--Congresses. 2. Morphology (Animals)--Congresses. I. Koppe, Thomas. II. Meyer, G. (Georg), 1948- III. Alt, Kurt W. IV. Title. V. Series: Frontiers of oral biology, v. 13. 1420-2433 ;
[DNLM: 1. Tooth--anatomy & histology--Congresses. 2. Dentition--Congresses. WU 101 I61c 2009]
QL858.I57 2008
599.9′43--dc22

2009029349

Bibliographic Indices. This publication is listed in bibliographic services, including Current Contents®.

www.karger.com
Printed in Switzerland on acid-free and non-aging paper (ISO 9706) by Reinhardt Druck, Basel
ISSN 1420–2433
ISBN 978–3–8055–9229–1
e-ISBN 978–3–8055–9230–7

This book is dedicated to the memory of three outstanding dental anthropologists:
Daris R. Swindler, Stanley M. Garn, and Coenraad F.A. Moorrees

Contents

Preface

Teeth and their supporting elements evolved as principal structures of mastication. Composed of distinct hard tissues and pulp, teeth are connected with the bony socket in which they are set by the periodontal ligament. Although all morphological features of the teeth and their surrounding structures are genetically determined to some extent, they are adapted to the environment in the course of ontogeny and phylogeny. Thus, teeth are exceptional sources for addressing important questions in numerous disciplines such as dental sciences, evolutionary biology, paleoanthropology, paleontology, archaeology, prehistoric anthropology, comparative anatomy, genetics, embryology, and forensic medicine. The last few decades have witnessed new and growing interest in dental morphology. This development in dental research was driven not only by the recent advances in data acquisition using highly sophisticated methods such as molecular analyses or new nondestructive imaging technologies, but also in the mode of data interpretation. Unfortunately, however, new insights into dental biology are usually presented at expert meetings or at several overlapping sessions of annual scientific conferences. Thus, it is more and more difficult to keep on track with new developments in this field.

To overcome these shortcomings, a group of enthusiastic scientists gathered in 1965 in Fredensborg, Denmark, to offer a forum for interdisciplinary communication and discussion in the field of dental morphology. This meeting was the beginning of a series of International Symposia on Dental Morphology that take place every 3 years. The Dental Morphology Symposia have always been most exciting meetings because of their special atmosphere and interdisciplinary character. Each symposium resulted in the publication of a volume with original papers. These papers, while presenting many details on several aspects of dental morphology, were intended to provide a broad picture on current aspects of research in dental morphology. Due to the interdisciplinary character, these volumes were not simply proceedings of scientific meetings. Although the information given in these volumes could not be complete, they provided the necessary background for further studies. Indeed, these volumes are still widely recognised and have a great impact on the further development of dental research.

The current volume contains a selection of papers that were presented at the 14th International Symposium on Dental Morphology held in Greifswald, Germany, August 27–30, 2008. Like the former volumes, the aim of this most recent addition to the series is to present progress in major topics of current research in dental morphology. In doing this, it is our hope that this volume will attract similar attention as the volumes of the former meetings. Presented in the current volume in the Karger series *Frontiers in Oral Biology*, the research presented at the 14th International Symposium on Dental Morphology has found the forum it deserves.

The 30 selected and fully refereed papers are arranged into six sections, including dental evolution, dental morphology, dental tissues, and dental growth and development. Due to recent advances in dental medicine, a section on clinical aspects of dental morphology is also included in this volume. A special feature of the present volume is the integration of new information about the role of teeth as tools in reconstructing the nature and behaviour of past populations in a section on teeth and reconstruction of the past.

To achieve a high standard in each section, outstanding scientists in their field were invited to act as co-editors. We are therefore most grateful to B. Holly Smith (Ann Arbor, Mich., USA), Mark F. Teaford (Baltimore, Md., USA), Inger Kjær (Copenhagen, Denmark), Alan Brook (Liverpool, UK), and John R. Lukacs (Eugene, Oreg., USA) for allocating the most suitable papers and for supervising the review process. We thank especially M. Christopher Dean (London, UK), for taking over the duties as a co-editor for the chapters on dental tissues from Moya M. Smith (London, UK). Due to health reasons, Moya M. Smith was not able to continue this work. Each section contains an in-depth introduction to the particular field of dental research. Thus, readers are encouraged to review these introductions in order to get the most out of the topics of this volume.

The preparation of the 14th International Symposium on Dental Morphology that let to this book was greatly supported by the Deutsche Forschungsgemeinschaft (DFG), the Deutsche Gesellschaft für Zahn-, Mund- und Kieferheilkunde, and the Medical Faculty of the Ernst Moritz Arndt University, Greifswald. We would like to thank Prof. Karlhans Endlich, head of the Department of Anatomy and Cell Biology of Greifswald University, and all members of his department. Without their direct and indirect help this symposium would not have been possible. In addition, we are most grateful to a number of individuals for their help and support including Dr. Frauke Fassbinder (Greifswald) and Priv.-Doz. Dr. Thomas Terberger (Greifswald). Finally, for their tremendous work during the meeting, we owe our special thank to the following medical and dental students of Greifswald University: Norman Apt, Torsten Bierdümpel, Hansgeorg Irmer, Sebastian Klug, Sandra Ortmann, Doreen Plaumann, Beate Roderer, Christoph Röth, Felix Rudolphi, Jessica Wickert and Eva-Maria Wittkowski.

Finally, we thank all the authors of this volume, not only for their speedy and efficient work, but especially for keeping their papers within the necessary space limitations. Last, but obviously not the least, it is our pleasure to acknowledge S. Karger publishers, Basel, Switzerland, for inviting this volume into their series *Frontiers in Oral Biology*, and the Karger team for their flexible and highly professional cooperation.

Thomas Koppe, Greifswald, Germany
Georg Meyer, Greifswald, Germany
Kurt W. Alt, Mainz, Germany

Previous volumes

1963 Dental Anthropology. DR Brothwell (ed). New York, Pergamon Press.

1967 Proceedings of the International Symposium on Dental Morphology. PO Pedersen, AA Dahlberg and V Alexandersen (eds). J Dent Res 46(suppl):769–992.

1971 Dental Morphology and Evolution. AA Dahlberg (ed). Chicago, University of Chicago Press.

1978 Development, Function and Evolution of Teeth. PM Butler and KA Joysey (eds). London, Academic Press.

1982 Teeth: Form, Function and Evolution. B. Kurtén (ed). New York, Columbia University Press.

1988 Teeth Revisited: Proceedings of the VIIth International Symposium on Dental Morphology. DE Russel, J-P Santoro, D Sigogneau-Russel (eds). Mémoires du Muséum National D'Histoire Naturelle, Sciences de la Terre, vol 53.

1992 Structure, Function and Evolution of Teeth. P Smith and E Tchernov (eds). London, Freund Publishing.

1995 Aspects of Dental Biology: Palaeontology, Anthropology and Evolution. J. Moggi-Cecchi (ed). Florence, International Institute for the Study of Man.

1995 Proceedings of the 10th International Symposium on Dental Morphology. RJ Radlanski and H Renz (eds). Berlin, M. Marketing Services.

1999 Dental Morphology 1998. JT Mayhall and T Heikkinen (eds). Finland, Oulu University Press.

2001 Dental Morphology 2001. A Brook (ed). Sheffield, Sheffield Academic Press.

2005 Current Trends in Dental Morphology Research. E Żądzińska (ed). Łódź, University of Łódź Press.

Koppe T, Meyer G, Alt KW (eds): Comparative Dental Morphology.
Front Oral Biol. Basel, Karger, 2009, vol 13, pp 1–2

Dental Evolution: An Introduction

B. Holly Smith

Museum of Anthropology, University of Michigan, Ann Arbor, Mich., USA

The proceedings of the First International Symposium on Dental Morphology appeared in the 1967 *Journal of Dental Research*. It was a classic, with papers by Brace, Brothwell, Butler, Dahlberg, Garn, Hiiemae, Koenigswald, Pedersen, Swindler and more. Those first proceedings represent a gathering of a great group of dental anthropologists, clinicians, morphologists and paleontologists who thought that representatives of all these different fields of knowledge about dentition should get together. Dahlberg, Pedersen and Alexandersen organized the first meeting in Fredensborg, Denmark in 1965, and since then the group has re-convened roughly every 3 years, typically in Europe. The 14th ISDM was held in Greifswald, Germany, August 27–30, 2008. Organized by Thomas Koppe (Greifswald), Kurt W. Alt (Mainz), and Georg Meyer (Greifswald), the meeting was a worthy successor to Al Dahlberg's vision. One hundred and forty-seven participants from 27 countries, including students and faculty, met over common interests. Contributions were well founded in the 1965 topics, but it was also evident that information preserved in teeth will soon breakthrough to depths unimagined in 1967. This volume has a sampling of papers presented in 2008.

An 'evolution' section could take many possible directions in terms of approaches, methods, taxa, and the recent or deep past. We lead with the broadest contribution. Jones gives an overview of the adaptive radiation of Rhynchochephalia into feeding niches that resulted in a diverse array of dentaries and skull form. The group, now largely extinct except for the living tuatara *(Sphenodon)*, was broadly distributed in the Mesozoic. Using this once diverse group, Jones shows how tooth form leads us to understand skull shape.

I recommend that anyone who studies mammalian dentition should spend some time in the Eocene. At that time, both modern and archaic orders coexisted, many with seemingly similar generalized molar patterns, especially among the herbivores. Dirks' group give us a look at two classic representatives of the extinct archaic ungulates *Phenacodus* and *Meniscotherium* (traditionally grouped as Condylarthra). With the use of incremental lines, they have begun to investigate the life histories of these archaic mammals. Their work begins what should eventually allow us to study the evolution of mammalian life histories and see whether mammal lives were structured as they are today in terms of relationships to body size and other variables.

Our remaining three papers take up anthropoid primates. Kupczik and coworkers break new ground by showing us the types of comparisons we can make with imaging technologies that allow us to visualize and measure roots, using it to show how these 'underpinnings' of teeth can differ in hard and soft object feeders. As they show for living anthropoids, there is a wealth of information within the jaws of fossil mammals.

Like Kupczik, Skinner and coworkers also use new imaging technologies to shapes that are either difficult to quantify or that would otherwise require destruction of the specimen. But here, their goal is to investigate discrimination of species rather than function. Remarkably, their analysis of the enamel-dentine junction successfully discriminated not only species of *Pan* but also subspecies within *Pan troglodytes.*

We arrive at hominins at the end of our section. Klinge's group use the material to investigate diagenesis, a general problem in paleontology, especially when more and more work in the future will be oriented toward retrieving signals of diet and life history from structures and chemical signatures in fossils. They find that even a seemingly poorly preserved specimen showed well-preserved microanatomy and apparent remnants of bacteria when observed with microradiography.

These five represent just part of the papers presented at the symposium, which also included two invited lectures. One is a synthess of mammal origins by Thomas Martin (Bonn, Germany) and the other a fascinating example of 'evo-devo' by Laurent Viriot (Poitiers, France) using mouse dentition. But the subject of evolution is so much a part of the frontiers of oral biology series that, in a way, we can claim all the papers in the volume as contributors to our knowledge of the evolution of dentition, and how dentition can inform us about evolution.

B. Holly Smith, PhD
Museum of Anthropology
University of Michigan
Ann Arbor, MI 48109-1107 (USA)
Tel. +1 734 662 5020, E-Mail bhsmith@umich.edu

Koppe T, Meyer G, Alt KW (eds): Comparative Dental Morphology.
Front Oral Biol. Basel, Karger, 2009, vol 13, pp 3–8

Phylogeny, Life History and the Timing of Molar Crown Formation in Two Archaic Ungulates, *Meniscotherium* and *Phenacodus* (Mammalia, 'Condylarthra')

Wendy Dirks[a] · Robert L. Anemone[b] · Patricia A. Holroyd[c] · Donald J. Reid[a] · Pamela Walton[a]

[a]School of Dental Sciences, Newcastle University, Newcastle upon Tyne, UK; [b]Department of Anthropology, Western Michigan University, Kalamazoo, Mich., and [c]Museum of Paleontology, University of California, Berkeley, Calif., USA

Abstract

The condylarths, or archaic ungulates, are a paraphyletic mammalian group including a number of fossil taxa whose relationships are unresolved. Included are two genera from the Paleocene and Eocene of North America, *Meniscotherium* and *Phenacodus*. Some workers place both genera in the family Phenacodontidae, while others exclude the highly dentally derived *Meniscotherium*. In this study, we use growth increments in histological thin sections to examine the timing of crown formation in five molars of *Meniscotherium* and one each of *Phenacodus intermedius* and *Phenacodus trilobatus*. We also use perikymata counts on an additional six molars of *Meniscotherium*. Although estimated body mass and molar dimensions in *Meniscotherium* are smaller than in either species of *Phenacodus*, molar formation times are longer, ranging from 0.71 to 1.44 years. Both *Phenacodus* molars take less than a year to form. Crown extension rates, the rate at which the crown grows in height, are as low as 3–15 μm per day in *Meniscotherium*, but range from 13 to 54 μm per day in *Phenacodus*. Although striae periodicities and daily enamel secretion rate are similar in both genera, the differences in the crown extension rate and overall timing of crown formation suggest differences in life histories and raise questions about the phylogenetic relationship of the two genera.

Growth patterns in the mammalian dentition reflect both phylogeny and life history characteristics and are key lines of evidence in understanding early mammal evolution, e.g. recent work by Asher and Lehmann [1] suggests Afrotheria may be characterized by unique patterns of tooth growth and eruption. However, data on dental growth in early mammals is generally lacking. The phenacodontid condylarths *Meniscotherium* (early Eocene) and *Phenacodus* (late Paleocene to early Eocene) are two taxa that are frequently included in analyses of afrotherian and/or mammalian evolution [2–4]. Here we present new data on dental growth in these two taxa that contributes to better understanding absolute and relative differences in growth and phylogenetic relationships within the Condylarthra.

Body mass in *M. chamense* is estimated to be 5–17 kg [5]. Both *P. intermedius* and *P. trilobatus* are estimated to be larger. Thewissen [6] estimated a body mass of 10–39 kg for *P. intermedius* and 22–87 kg for *P. trilobatus*. Brain size, however, is relatively larger in *Meniscotherium* [7, 8]. Both species of *Phenacodus* have larger molars than *Meniscotherium.* Given the larger body and molar size in *Phenacodus*, we might predict that *Phenacodus* molars will take longer to form than *Meniscotherium* molars. Molar formation time in primates is correlated with both body and brain size [9], however, and the relatively larger brain size in *Meniscotherium* could be correlated with slower molar formation times.

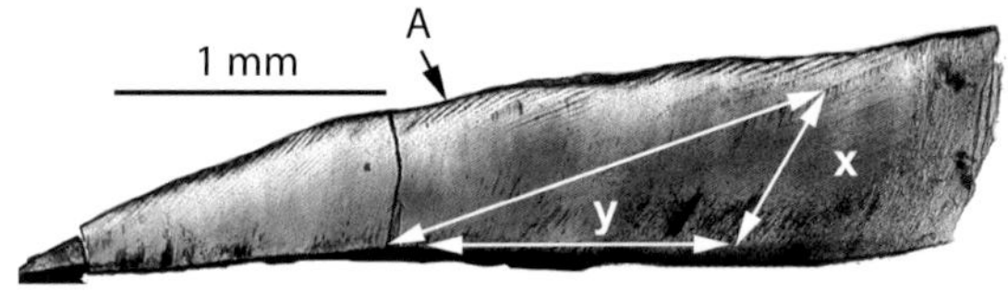

Fig. 1. Lateral enamel of *Meniscotherium chamense* upper molar fragment illustrating the method used to calculate enamel extension rate. Line x represents the distance measured along a prism to a prominent accentuated line. Line y represents the distance between the intersection of the accentuated line and line x with the enamel dentine junction. Arrow A points to striae of Retzius reaching the enamel surface as perikymata.

Methods

The histological study comprised sections from one upper right molar of *P. trilobatus*, one lower left molar of *P. intermedius*, and two left M_1s, one left M_2, a lower left molar fragment and a premolar fragment of *M. chamense.* Perikymata counts were done using a Wild Stereomicroscope from six additional *M. chamense* molars from a left maxillary fragment with M^{2-3}, a right maxillary fragment with M^{1-3}, and an isolated left M_3. All specimens are from the Wasatchian NALMA.

Histological sections prepared to a thickness of approximately 100 μm were analyzed using an Olympus BX51 microscope mounted with a Q-Imaging Micropublisher 3.3 RTV camera and Improvision Openlab 5.0.2 image analysis software. In enamel, daily growth increments are called cross-striations [10, 11] and longer period increments are called striae of Retzius [12, 13]. These growth increments were used to determine daily secretion rate (DSR), striae periodicity, and crown extension rate.

Two methods were used to determine DSR. In some sections, the distance between adjacent cross striations was directly measured along the enamel prisms. In other sections, two points were marked along a prism and the length between them measured while simultaneously counting the number of cross striations. DSR was determined by dividing the length by the number of cross striations, generally less than ten. To determine whether there were differences between genera, DSR was compared in the inner, mid, and outer cuspal enamel in the protoconid and metaconid in the *P. intermedius* molar and a *Meniscotherium* M_1. Similar measurements were made in the cuspal enamel of the *P. trilobatus* molar protocone and a *Meniscotherium* upper molar fragment.

The number of days between striae of Retzius (striae periodicity) is invariant between teeth in an individual [11, 14]. Striae periodicity was determined by direct count of cross-striations between adjacent striae or dividing the distance between adjacent striae by the DSR in that area.

The crown extension rate is the rate at which the crown increases in height each day and is determined by the size of the cohort of differentiating ameloblasts along the inner enamel epithelium [15]. The method for determining extension rate is illustrated in figure 1 and is modified from the methods of Shellis [15], Reid et al. [16] and Dean and Vesey [17]. The distance along an enamel prism was measured from the enamel-dentine junction (EDJ) to its intersection with a prominent accentuated line or stria of Retzius. That distance was then divided by the average DSR along the prism to yield the number of days (x) required to form that length of prism. The accentuated line, representing the forming front of enamel after x days, was then followed back to its intersection with the EDJ and the length (y) of the EDJ measured between the two intersection points. Length y represents the increase in crown height in x days and daily extension rate equals y/x.

Crown formation times were determined in three ways. In the histological sample, cuspal enamel formation time in days was determined by measuring the distance along a prism from the EDJ to the point where the most occlusal striae of Retzius cropped out at the enamel surface. This distance was divided by the average DSR along that prism. In *P. trilobatus*, it was impossible to count all of the striae of Retzius and crown formation time was determined at the same time and in the same manner as that used in determining crown extension rate [16]. Beginning in the cuspal enamel, DSR, prism length and the number of days to the formation of an accentuated line were determined. The line was followed back to the EDJ and at the point of intersection, DSR, prism length, and number of days to form enamel to another accentuated line were determined. That line was followed back to the EDJ and the process repeated until the timing for formation of the entire height of the crown was determined. In *P. intermedius* and *M. chamense*, lateral enamel formation time in days was determined by multiplying the number of striae of Retzius by the striae periodicity. The total crown formation time was determined by adding cuspal and lateral formation time together. Lateral enamel formation time in *M. chamense* was also estimated in a larger sample using perikymata counts by multiplying the number of perikymata in each molar cusp by three different striae periodicities determined from the histological sample. In those cusps for which cuspal enamel formation times had been derived histologically, a total estimated crown formation time was determined by adding the lateral formation time based on perikymata counts.

Results

Striae periodicities for *Phenacodus trilobatus* and *P. intermedius* were five and six, respectively. One of the *Meniscotherium chamense* teeth had a striae periodicity of four and two teeth had a periodicity of five. In two other teeth, it was impossible to determine the periodicity precisely. In one tooth, it appeared to be either four or five, and in another, either five or six. In determining lateral enamel formation time in the perikymata counts, periodicities of four, five and six were used. Despite differences in body mass, striae periodicity is similar in both *Phenacodus* species and *Meniscotherium*.

DSR was similar in all the *Meniscotherium* molars, ranging from 2.8 to 3.6 μm in cuspal enamel and 3.1 to 4.6 μm in lateral enamel. There appear to be very few differences in DSR between species. In the protocone of the *P. trilobatus* upper molar, DSR was 2.6 μm in the inner enamel, increasing gradually to 3.5 μm in the mid enamel and 4.4 μm in the outer enamel. In the protocone of the *M. chamense* upper molar fragment, DSR was 3.9 μm in inner enamel, decreasing to 3.2 μm in the mid enamel and then increasing to 4.0 μm in the outer enamel. In the protoconid and metaconid of the *P. intermedius* lower molar, DSR was 2.8 μm in the inner enamel, 3.3 μm in mid enamel and 3.8 μm in outer enamel. DSR was similar in the protoconid and metaconid of a *M. chamense* M_1, ranging from 2.9 μm in the inner enamel, and increasing to 3.5 μm in both mid and outer enamel.

Crown formation times calculated from histological sections are given in table 1. Lateral formation time from perikymata counts is given in table 2, along with estimated minimum and maximum crown formation times when cuspal enamel formation for the appropriate cusp type is added from the histological sections. Cusp nomenclature is from Williamson and Lucas [5]. Despite similar DSR, *Meniscotherium* molars consistently take longer to form than molars from the larger *Phenacodus* species.

Crown extension rate is higher in both *Phenacodus* species than in *Meniscotherium*, but in both genera, crown extension rate is higher in mandibular than in maxillary molars. In the *M. chamense* M_1, the extension rate is 15 μm per day in the first 1.5 mm of crown height, and then decreases to 8.9 μm per day when crown height reaches 2.3 mm. In the upper molar fragment, the extension rate is lower, decreasing from 4.7 to 3.3 μm per day. In the *P. intermedius* lower molar, the extension rate is extremely rapid, reaching 54.7 μm per day in the cervical region. The extension rate in the *P. trilobatus* upper molar is 20.7 μm per day in the first 2.7 mm, decreasing to 13.3 μm per day when crown height reaches 4.7 mm. It increases again in the next 0.2 mm, to 23.9 μm per day. This anomalous decrease occurs along the EDJ at a point under a severe linear enamel hypoplasia at the enamel surface, suggesting that it is a pathological response to a severe stress during tooth development rather than something typical of the species.

Table 1. Crown formation time in *Meniscotherium* and *Phenacodus* from histological sections

Species	Specimen	Locality	Tooth	Cusp	Cuspal days	Lateral days	Total days	Total years
M. chamense	WMUVP 5180	148, Wasatch Fm, Great Divide Basin, Wyoming	LM_1	protoconid	132	255	387	1.06
				metaconid	147	175	322	0.88
				hypoconid	154	225	379	1.04
				entoconid	104	155	249	0.71
M. chamense	uncatalogued	Great Divide Basin, Wyoming	upper molar fragment	protocone	177	350	527	1.44
P. intermedius	UCMP 39870L	V5029, Willwood Fm, Bighorn Basin, Wyoming	lower molar	protoconid	132	156	288	0.79
				metaconid	126	144	270	0.74
P. trilobatus	UCMP 39870U	V5029, Willwood Fm, Bighorn Basin, Wyoming	upper molar	protocone	–	–	345	0.95

WMUVP = Western Michigan University Vertebrate Paleontology; UCMP = University of California Museum of Paleontology.

Conclusions

Despite larger body and molar size in *Phenacodus*, *Meniscotherium* molars take longer to form and grow more slowly in height. One possible explanation for this is the relatively larger brain in *Meniscotherium*. The encephalization quotient (EQ), is a measure of brain size relative to body mass based on Jerison's [18] allometric equation $E = 0.12P^{0.67}$, where E is predicted brain size and P is body mass in grams. EQ in *Meniscotherium* is estimated as 0.14–0.37 but only 0.18–0.20 in *Phenacodus* [5, 18, 19]. In primates, molar formation times are correlated with both brain and body size [9]. In primates and other mammals, many aspects of dental development are correlated with the pace of life history [20–24], which is also correlated with brain size. It is possible that *Meniscotherium* had a slower life history than did *Phenacodus*. These results also demonstrate considerable interspecific variability in growth rates that may bear on questions of phylogeny. Further work on other aspects of dental development is required to resolve this question, but these results highlight the promise of this line of inquiry to clarifying questions about early mammal diversification.

Acknowledgment

We thank Tom Williamson, Spencer Lucas, and Justin Spielmann of the New Mexico Museum of Natural History and Science for access to specimens.

Table 2. Crown formation (cf) time in *Meniscotherium* from perikymata counts and histologically derived cuspal formation times: all specimens from Locality 203, San Jose Formation, San Juan Basin, New Mexico

Specimen	Tooth	Cusp	Total Perikymata	Lateral CF days, periodicity 4	Lateral CF days, periodicity 5	Lateral CF days, periodicity 6	Cuspal days	Estimated minimum and maximum total crown formation times in years
NMMNH P3467	M^1	paracone	57	228	285	342	–	–
		metacone	60	240	300	360	–	–
		protocone	41	164	205	246	177	0.93–1.16
		hypocone	35	140	175	210	–	–
	M^2	paracone	63	252	315	378	–	–
		metacone	67	268	335	402	–	–
	M^3	paracone	52	208	260	312	–	–
		metacone	67	268	335	402	–	–
		protocone	46	184	230	276	177	0.99–1.24
NMMNH P3334	M^2	paracone	78	312	390	468	–	–
		metacone	77	308	385	462	–	–
		protocone	52	208	260	312	177	1.05–1.34
		hypocone	40	160	200	240	–	–
	M^3	paracone	70	280	350	420	–	–
		metacone	64	256	320	384	–	–
		protocone	57	228	285	342	177	1.11–1.42
NMMNH P3249	M_3	metaconid	28	112	140	168	–	–
		entoconulid	45	180	225	270	–	–
		entoconid	44	176	220	264	104	0.77–1.05

NMMNH = New Mexico Museum of Natural History and Science.

References

1 Asher RJ, Lehmann T: Dental eruption in afrotherian mammals. BMC Biol 2008;6:14.
2 Asher R: A web-database of mammalian morphology and a reanalysis of placental phylogeny. BMC Evol Biol 2007;7:108.
3 Asher RJ, Novacek MJ, Geisler JH: Relationships of endemic African mammals and their fossil relatives based on morphological and molecular evidence. J Mammal Evol 2003;10:131–194.
4 Thewissen JGM, Domning DP: The role of phenacodontids in the origin of the modern orders of ungulate mammals. J Vert Paleo 1992;12:494–504.
5 Williamson TE, Lucas SG: *Meniscotherium* (Mammalia, 'Condylarthra') from the Paleocene-Eocene of Western North America. Bull New Mex Mus Nat Hist Sci 1992;1:1–75.
6 Thewissen JGM: Evolution of Paleocene and Eocene Phenacodontidae (Mammalia, Condylarthra). U Mich Papers Paleo 1990;29:1–107.
7 Cope ED: The Vertebrata of the Tertiary Formations of the West. Book 1. Report of the US Geological Survey of the Territories (Hayden) 1884, vol 3, pp 1–1009.
8 Gazin CL: A study of the early Tertiary condylarthran mammal *Meniscotherium*. Smithsonian Misc Coll 1965;149:1–98.
9 Macho GA: Primate molar crown formation times and life history evolution revisited. Am J Primatol 2001;55:189–201.

10 Boyde A: The structure of developing mammalian dental enamel; in Stack MR, Fearnhead RW (eds): Tooth Enamel: Its Composition, Properties, and Fundamental Structure. Bristol, John Wright & Sons, 1965, pp 163–167.

11 Smith TM: Experimental determination of the periodicity of incremental features in enamel. J Anat 2006;208:99–113.

12 Retzius A: Bemerkungen über den inneren Bau der Zähne, mit besonderer Rücksicht auf dem in Zahnknochen vorkommenden Röhrenbau. Arch Anat Physiol 1837:486–566.

13 Dean MC: Incremental markings in enamel and dentine: what they can tell us about the way teeth grow; in Teaford MF, Smith MM, Ferguson MWJ (eds): Development, Function and Evolution of Teeth. Cambridge, Cambridge University Press, 2000, pp 119–130.

14 FitzGerald CM: Do enamel microstructures have regular time dependency? Conclusions from the literature and a large-scale study. J Hum Evol 1998;35:371–386.

15 Shellis RP: Variations in growth of the enamel crown in human teeth and a possible relationship between growth and enamel structure. Archs Oral Biol 1984;29:697–705.

16 Reid DJ, Schwartz GT, Chandrasekera MS, Dean MC: A histological reconstruction of dental development in the common chimpanzee, *Pan troglodytes*. J Hum Evol 1998;35:427–448.

17 Dean MC, Vesey P: Preliminary observations on increasing root length during the eruptive phase of tooth development in modern humans and great apes. J Hum Evol 2008;54:258–271.

18 Jerison HJ: Evolution of the Brain and Intelligence. New York, Academic Press, 1973.

19 Radinsky L: Evolution of brain size in carnivores and ungulates. Am Nat 1978;112:815–831.

20 Smith BH, Crummett TL, Brandt KL: Ages of eruption of primate teeth: a compendium for aging individuals and comparing life histories. Yrbk Phys Anthropol 1994;37:177–231.

21 Smith BH: 'Schultz's rule' and the evolution of tooth emergence and replacement patterns in primates and ungulates; in Teaford MF, Smith MM, Ferguson MWJ (eds): Development, Function and Evolution of Teeth. Cambridge, Cambridge University Press, 2000, pp 212–227.

22 Dean C, Leakey MG, Reid DJ, Schrenk F, Schwartz GT, Stringer C, Walker A: Growth processes in teeth distinguish modern humans from *Homo erectus* and earlier hominins. Nature 2001;414:628–631.

23 Dean MC: Tooth microstructure tracks the pace of human life-history evolution. Proc Royal Soc [B] 2006;273:2799–3808.

24 Dirks W, Bowman JE: Life history theory and dental development in four species of catarrhine primates. J Hum Evol 2007;53:309–320.

Wendy Dirks, PhD
Oral Biology, School of Dental Sciences
Framlington Place
Newcastle upon Tyne NE2 4BW (UK)
Tel. +44 0 191 222 7918, Fax +44 0 191 222 6137, E-Mail Wendy.Dirks@ncl.ac.uk

Koppe T, Meyer G, Alt KW (eds): Comparative Dental Morphology.
Front Oral Biol. Basel, Karger, 2009, vol 13, pp 9–15

Dentary Tooth Shape in *Sphenodon* and Its Fossil Relatives (Diapsida: Lepidosauria: Rhynchocephalia)

M.E.H. Jones

Research Department of Cell and Developmental Biology, UCL, University College London, London, UK

Abstract

Background: Today Rhynchocephalia, the sister taxon to Squamata (snakes, lizards and amphisbaenians), is only represented by the tuatara *(Sphenodon)* of New Zealand. However, for much of the Mesozoic, the group was speciose and globally distributed. Historically, the Rhynchocephalia were considered to be homogenous and unspecialized but new fossils and new research are overturning this view. As well as differences in body size, body proportions, habit (aquatic vs. terrestrial), and skull structure, their teeth show variation in shape, size, number, arrangement and enamel thickness. This suggests differences in diet and mode of feeding. The teeth of basal taxa tend to be relatively simple and conical, whereas those of derived taxa possess complex flanges and wear facets. **Methods:** Dimensions of the dentary tooth bases were measured in apical view for a large sample of rhynchocephalian taxa. **Results:** These measurements reveal three general tooth types: small ovoid teeth, large wide teeth, and large elongate teeth. **Conclusion:** These three categories correspond to food processing as inferred from tooth wear (puncturing+crushing, grinding+shredding and tearing+cutting, respectively). A phylogenetic signal is also present as the teeth of basal taxa generally conform to the first category. The larger tooth bases of derived taxa provide stronger attachment and contribute to a stouter tooth shape more resistant to loading and torsional forces. This in turn corresponds to skull architecture because the skulls of derived taxa could accommodate larger jaw muscles with a greater leverage relative to basal taxa.

Rhynchocephalia is a group of diapsid reptiles that separated from its sister clade Squamata (lizards, amphisbaenians, snakes) about 240–250 mya and radiated worldwide during the Early-Mid Mesozoic (230–175 mya) [1, 2]. Towards the end of the Mesozoic (175–65 mya), the known distribution of the group became increasingly restricted to southern continents [1]. Little is known of their subsequent history apart from a relatively recent fossil record from New Zealand (19–0 mya) where the only living member, *Sphenodon* (the tuatara), now resides.

For a long time the Rhynchocephalia were considered to be conservative with respect to many aspects of their anatomy, including their teeth [3]. However, work over the last three decades and discovery of new fossil material has challenged this view. We now know of six broad subgroups: a paraphyletic series of 'basal taxa' (small gracile forms), clevosaurs (small robust forms), pleurosaurs (long bodied aquatic forms), sphenodontines (*Sphenodon* and certain poorly

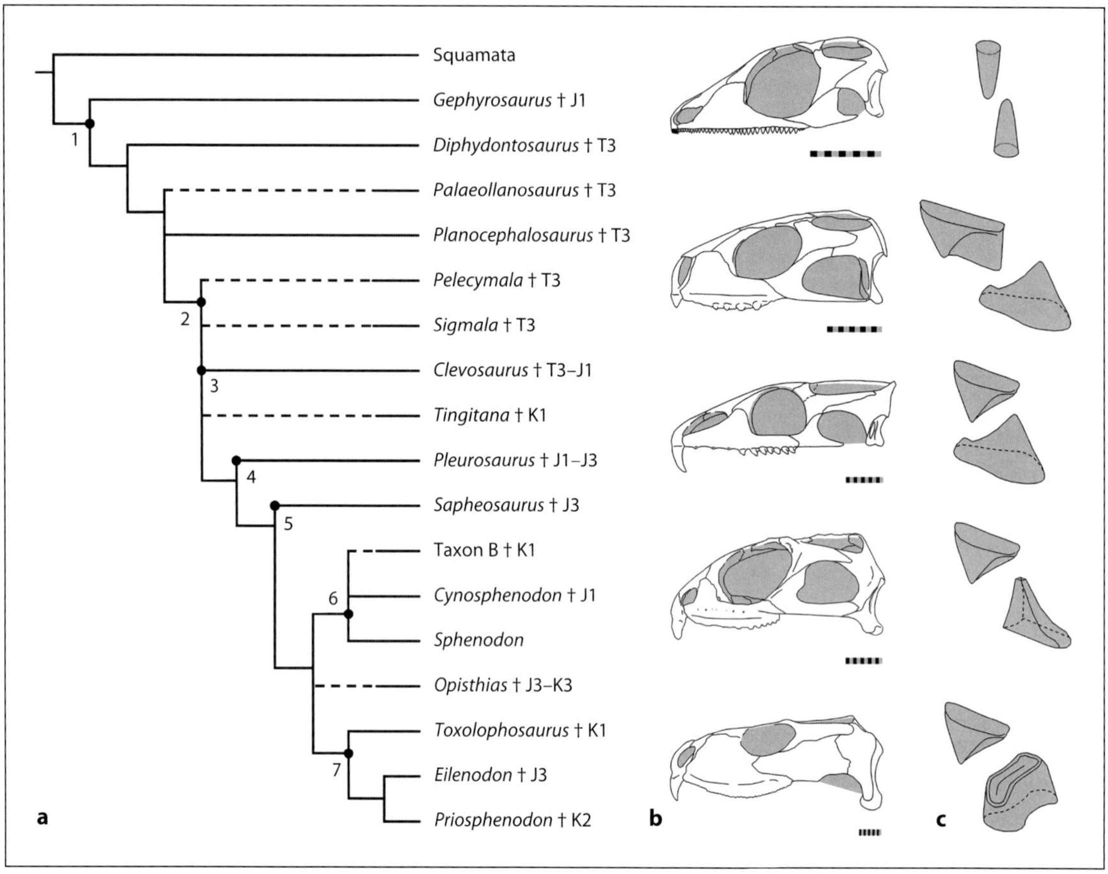

Fig. 1. **a** Phylogenetic hypothesis for selected Rhynchocephalia (based on data from [1, 2, 5, 7, 12, 18–20, 22]). Dotted lines indicate unresolved relationships. Nodes: 1, Rhynchocephalia; 2, derived taxa; 3, clevosaurs; 4, pleurosaurs; 5, sapheosaurs; 6, sphenodontines; 7, eilenodontines. Known time distribution: J1, Early Jurassic; J3, Late Jurassic; K1, Early Cretaceous; K2, Late Cretaceous; T3, Late Triassic. **b** Skulls of *Gephyrosaurus, Clevosaurus, Palaeopleurosaurus, Sphenodon, and Priosphenodon* (redrawn from Jones [4]), scale = 10 mm. c Diagrammatic representations of left maxillary and dentary teeth in posterolateral view for basal taxa, clevosaurs, pleurosaurs, sphenodontines and eilenodontines.

known Mesozoic taxa), and eilenodontines (relatively large heavily built forms) [4] (fig. 1). There are also taxa that either fall outside these groupings or are of uncertain affinity. As well as variation in body size, body shape and habit (aquatic vs. terrestrial), rhynchocephalians demonstrate substantial variation in skull structure and in the size, number, shape and arrangement of the teeth.

Basal taxa (*Gephyrosaurus, Diphydontosaurus, Planocephalosaurus* [5–7]) possess many (12–40) simple teeth along their jaws and palate, with the lateral most palatine tooth row enlarged, as in all known Rhynchocephalia [2]. The elongate articular surface of the lower jaw suggests some propalinal jaw motion [5–7]. Clevosaurs possess a few (<5) large blade-like marginal teeth and a reduced number of teeth on their palate [6, 8].

Conspicuous wear facets indicate a strong scissor-like orthal shearing action. The teeth of the aquatic sapheosaurs and pleurosaurs are generally considered to be blade-like and somewhat similar to those of clevosaurs [9]. Eilendontines possess teeth with thickened enamel and well-defined wear facets from propalinal shearing [10–12]. Maxillary teeth bear posterolingual flanges whereas those on the palatine and dentary are transversely expanded. The modern *Sphenodon* possesses teeth of a similar shape to the maxillary teeth of eilenodontines but the palatine teeth are not expanded and the dentary teeth are more elongate and pyramidal with small anterior flanges [13, 14]. *Sphenodon* jaw movements comprise an anteriorly directed shearing phase (prooral propaliny) following almost complete jaw closure [14].

The diversity in tooth morphology demonstrates that Mesozoic rhynchocephalians were adapted to exploit a range of different food resources [1, 6, 11, 15]. *Sphenodon* feeds on a wide variety of food items but mainly arthropods such as large beetles [4, 16]. Basal taxa possess teeth ideally suited for piercing small insects [5–7, 17]. The diet of clevosaurs remains uncertain, but may have been flexible like that of *Sphenodon* [6, 8, 4, 15]. Because of their robust jaws, extensive wear and thickened enamel, eilenodontines are considered to have been herbivorous [4, 11].

Here I revisit the study by Jones [15] with an expanded data set.

Material and Method

In total, 546 teeth from the lower jaw were examined, more than double the sample size used previously [15]. The data set includes further specimens of *Clevosaurus*, *Sphenodon*, *Eilenodon* and *Planocephalosaurus*. Unfortunately, adequate dentary teeth of sapheosaurs and pleurosaurs were unavailable for study. Otherwise the sample is fairly balanced with a large number of teeth representing each of the remaining major phylogenetic groupings as well as some taxa of uncertain affinity such as *Paleollanosaurus* [18] and *C. latidens* [19]. The anterior hatchling teeth and caniniform teeth of *Sphenodon* and *Cynosphenodon* will be dealt with elsewhere.

The teeth were drawn in apical/occlusal view using a Wild Stereo Microscope and *camera lucida* after the lower jaw was orientated in a standardized position. When lower jaws has a slightly concave dorsal margins (such as those of *Sphenodon*), three or four teeth were drawn at a time before reorientation. Images of teeth from published descriptions were used for *Clevosaurus convallis*, *Cynosphenodon*, *Opisthias*, *Paleollanosaurus*, *Pelecymala*, *Sigmala*, Taxon B, *Tingitana*, and *Toxolophosaurus* [11, 18–22].

For each tooth two dimensions were measured: length (greatest mesiodistal dimension parallel to the labial margin of the jaw) and width (greatest labiolingual dimension perpendicular to the margin of the jaw) (fig. 2). Invariably, the cusp tapers up from the base, so these measurements correspond to the dimensions of the tooth base. In Rhynchocephalia, teeth are not usually replaced; instead new teeth are added to the rear of the tooth row during growth of the jaw bone [13, 20]. Therefore, anterior teeth are often more worn and may plot differently because of differences in lateral or lingual wear rather than their original shape [15]. However, in many rhynchocephalian taxa, worn teeth are present for much of an individual's life (e.g. *Clevosaurus*, *Eilenodon*, *Sphenodon* and *Toxolophosaurus* [8, 10, 11, 13]). Thus, the shape of a tooth following wear is the 'effective tooth shape' and part of the animal's feeding apparatus (e.g. [8, 11]).

Inferred tooth function is based on careful examination of wear facets and the finer details of tooth shape.

Results

The teeth of most basal taxa *(Gephyrosaurus, Diphydontosaurus)* have small ovoid tooth bases with similar mesiodistal and labiolingual dimensions (fig. 3). Derived rhynchocephalians instead tend to have teeth with larger tooth bases but with sub-equal proportions. All teeth belonging to clevosaurs (e.g. *Clevosaurus convallis*, *C. hudsoni*) and most belonging to sphenodontines *(Cynosphenodon, Sphenodon)* plot as together as elongate (labiolingual dimension>mesiodistal dimension) with many being twice as long as wide. In contrast, eilenodontines *(Eilenodon, Toxolophosaurus)* usually plot as wider than long (mesiodistal dimension>labiolingual dimension).

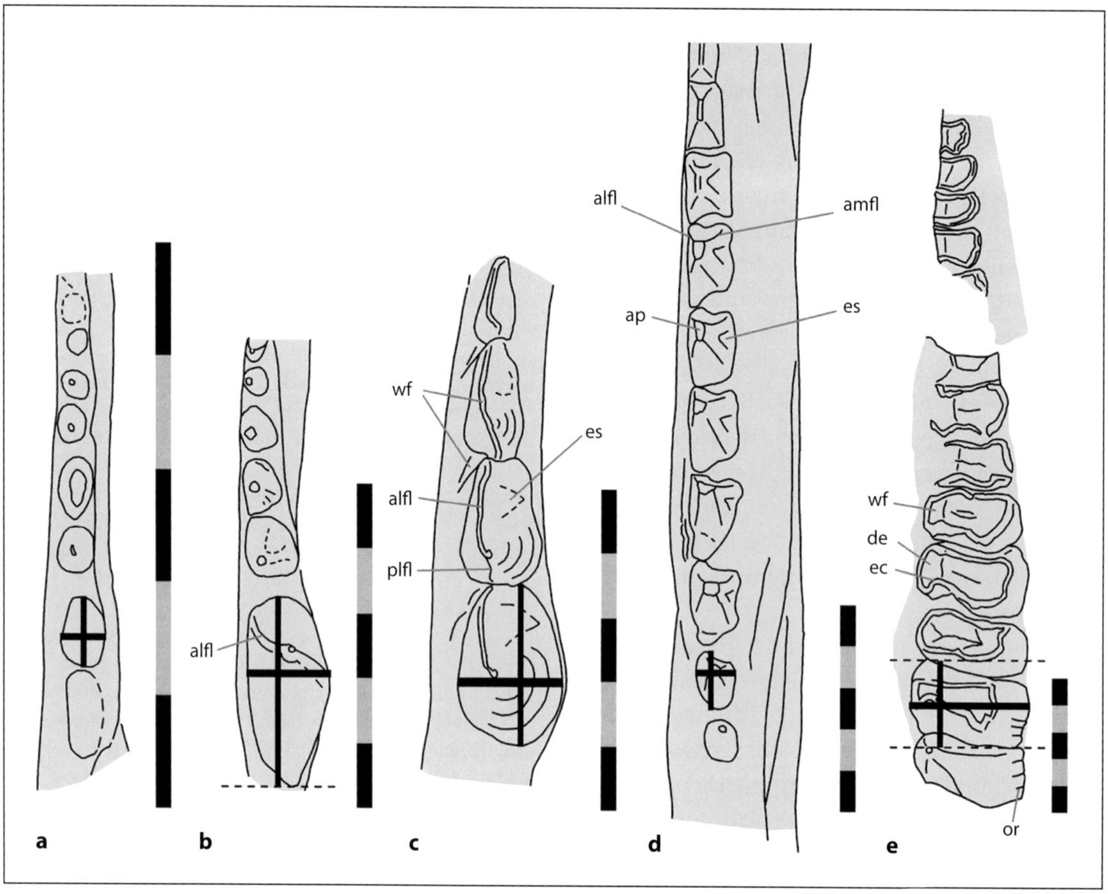

Fig. 2. Drawings of rhynchocephalian lower jaws in apical view. **a** *Diphydontosaurus.* **b** *Planocephalosaurus.* c *Clevosaurus.* **d** *Sphenodon* specimen LDUCZ x804. **e** *Eilenodon* DMNH 10685. Black bars represent examples of length and width measurements taken. ap = Apex; afl = anterior flange; alfl = anterolateral flange; amfl = anteromedial flange; de = dentine; ec = enamel crest; es = escape structure; or = ornament/escape structure; pfl = posterior flange; wf = wear facet. All are left jaws except for **b**, which is reversed for comparison. Scale bars = 5 mm.

They can be a similar length to those of sphenodontines or clevosaurs (~2 mm), but the width can be twice to five times wider (5 mm+).

The small teeth of *Paleollanosaurus* [18] and Taxon B [22] plot amongst the basal taxa, whereas those of *C. latidens* [19], *Opisthias* [11] and *Pelecymala* [20] are wider than long and plot amongst the smallest (usually anteriorly positioned and highly worn) eilenodontine teeth. The dentary teeth of *Sigmala* [20] plot in various positions but mainly within the elongate group. *Planocephalosaurus* [6] teeth are mostly small and ovoid but the large flanged posteriormost tooth plots within the area otherwise occupied by *Sphenodon* and *Clevosaurus. Cynosphenodon* [21] and *Tingitana* [22] plot with similar aspect ratios to clevosaurs and *Sphenodon* but their distributions do not greatly overlap because they are smaller.

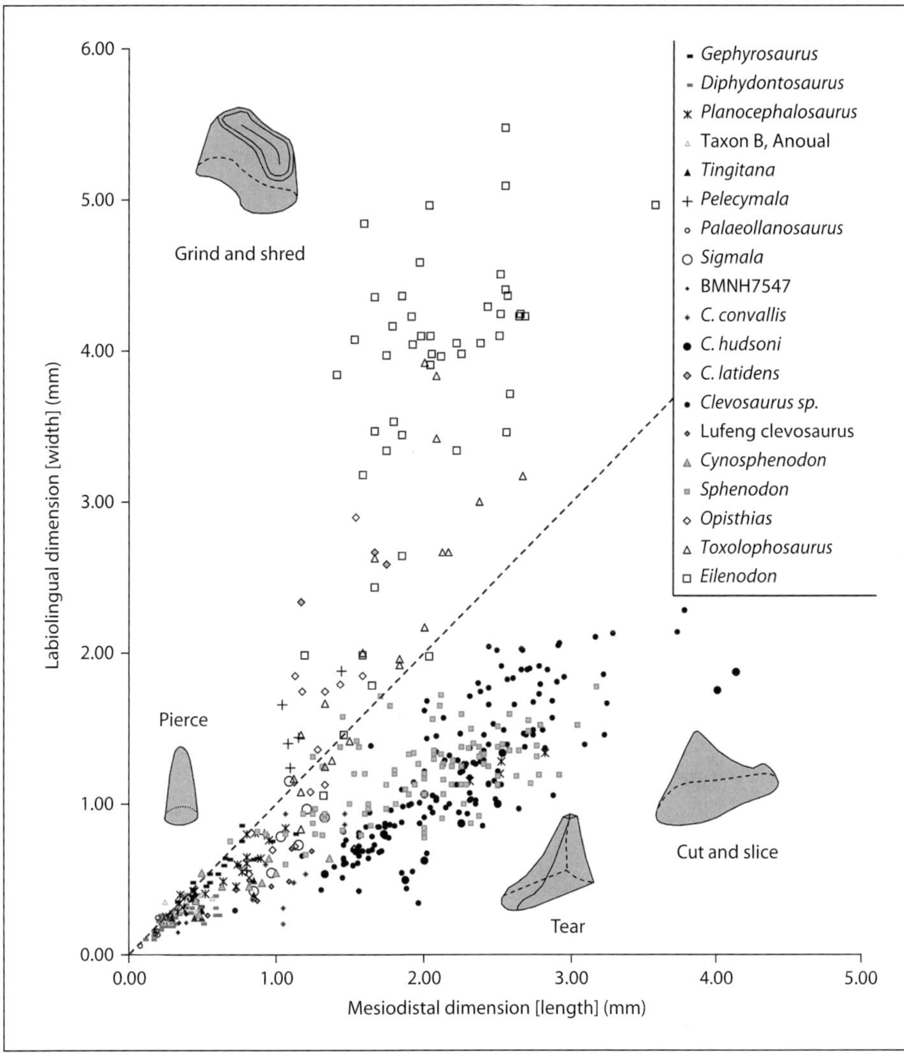

Fig. 3. Dimensions of the dentary tooth base in Rhynchocephalia. Dashed line represents values of equal width and length. Diagrammatic representations of right dentary teeth in posterolateral view are positioned near their respective data points.

Discussion

Overall, the additional data is consistent with previous findings [15] that basic tooth dimensions (mesiodistal dimension vs. labiolingual dimension) are able to define three broad morphotypes among rhynchocephalian dentary teeth: small columnar teeth with ovoid bases, large wide teeth, and large elongate teeth. The first morphotype contains teeth that are suited to efficient puncturing and crushing [17]. The second and third types correspond to the two alternate modes of food reduction described by Lucas and Luke [4, 23]; shredding and grinding or tearing and slicing.

A large tooth base width, for example in eilenodontines, increases the potential surface area of contact with food items [23]. Assuming even jaw closure and regular tooth height, this may reduce the relative amount of point loading possible. However, it means that anterior and posterior transverse blades caused by wear on the thickened enamel can occur in large numbers, as in modern mammals [11]. This tooth type is mainly restricted to the second half of the Mesozoic when the distribution of Rhynchocephalia was apparently contracting. Two exceptions to this are the Late Triassic *C. latidens* and *Pelecymala* [19, 20]. The current phylogenetic positions of these taxa suggest that a wide tooth type evolved at least twice within Rhynchocephalia. However, both Triassic taxa are known from very limited material. *Pelecymala* lacks the dorsal wear facets of eilenodontines and is described as using orthal shearing rather than propaliny [19], *C. latidens*, however, does possess the concave anterior and posterior surfaces of eilenodontine teeth.

The long tooth base of clevosaurs permit a series of long blades arranged near parallel with the margin of the jaw orientated in order to cut effectively against blades on the maxillary teeth during orthal shearing [24]. The small number of teeth permitted limits the amount of food that can be worked upon but initial point loading is high [23, 24]. The long tooth bases of sphenodontines permit several cutting edges in more than one orientation. Posterior and dorsal edges cut during jaw closure whereas anterior flanges improve food item penetration during jaw closure [25] and also cut during prooral shearing [14].

As tooth implantation in Rhynchocephalia is generally acrodont (fused to the crest of the jaw bone) and the crown tapers from the tooth base, the dimensions measured provide some indication as to the surface area of attachment. The teeth of derived taxa have larger bases than those of basal taxa and this contributes to their stout shape. Correspondingly, increased muscle capacity, shorter jaws (out levers), and reduced fenestration suggest that derived taxa were capable of harder bites than basal taxa [4]. There is some positive relationship between tooth size and skull size (skull length). For example, the teeth with the largest tooth base dimensions are found in eilenodontines (skull length = 60–110 mm) [11, 12] and the teeth with the smallest base dimensions are found in *Diphydontosaurus* (skull length = 10–15 mm [7]). However, the posterior tooth in *Clevosaurus* (skull length up to 40 mm long [8]) can be both longer and wider than teeth found in *Sphenodon* (skulls sampled: 43–67 mm). Similarly, the posteriormost tooth in *Planocephalosaurus* (skull length = 20 mm [6]) can have bases as large as those of teeth in *Sphenodon*.

Rhynchocephalia demonstrate substantial diversity in tooth structure and counter perceptions that the teeth of all reptiles are simple or homogenous.

Acknowledgements

For access to material, I thank the staff at Grant Museum of Zoology UCL (UK), University Museum of Zoology Cambridge (UK), The Natural History Museum (UK), Denver Museum of Natural History (USA), Natural History Museum of Los Angeles (USA), and the Institute of Vertebrate Palaeontology and Palaeoanthropology (China). Most of this work was carried out during a PhD funded by the BBSRC and supervised by Professor Susan E. Evans (UCL) who I also thank, along with my PhD examiners Dr Emily Rayfield (University of Bristol) and Professor J. Chris Buckland-Wright (Guy's, Kings College, University of London), and one anonymous reviewer, for constructive comments on an earlier version of this work. Funding UCL Graduate School and a Royal Society Chinese Academy of Sciences Joint Project grant (awarded to S.E. Evans and Dr. Yuan Wang [IVPP]) assisted with travel expenses.

References

1 Evans SE, Prasad GVR, Manhas BK: Rhynchocephalians (Diapsida: Lepidosauria) from the Jurassic Kota Formation of India. Zool J Linn Soc 2001;133:309–334.

2 Evans SE: At the feet of dinosaurs: the early history and radiation of lizards. Biol Rev 2003;78:513–551.

3 Edmund AG: Dentition; in Gans C, Bellairs A d'A, Parsons TS (eds): Biology of the reptilia. I. Morphology. New York, Academic Press, 1969, pp 117–200.

4 Jones MEH: Skull shape and feeding strategy in *Sphenodon* and other Rhynchocephalia (Diapsida: Lepidosauria). J Morphol 2008;269:945–966.

5 Evans SE: The skull of a new eosuchian reptile from the Lower Jurassic of South Wales. Zool J Linn Soc 1980;70:203–264.

6 Fraser NC, Walkden GM: The ecology of a Late Triassic reptile assemblage from Glouchestershire, England. *Palaeogeog Palaeoclim Palaeolecol* 1983;42:341–365.

7 Whiteside DI: The head skeleton of the Rhaetian sphenodontid *Diphydontosaurus avonis gen. et sp. nov.*, and the modernising of a living fossil. Phil Trans R Soc 1986;B312:379–430.

8 Fraser NC: The osteology and relationships of *Clevosaurus* (Reptilia: Sphenodontida). Phil Trans R Soc B 1988;321:125–178.

9 Cocude-Michel M: Les rhynchocéphales et les sauriens des calcaires lithographiques (Jurassique Supérieur) d'Europe occidentalle. Nouv Arch Mus Hist Nat Lyon 1963;7:1–187.

10 Rasmussen TE, Callison G: A new herbivorous sphenodontid (Rhynchocephalia: Reptilia) from the Jurassic of Colorado. J Paleontol 1981;55:1109–1116.

11 Throckmorton GS, Hopson JA, Parks P: A redescription of *Toxolophosaurus cloudi* Olson, a Lower Cretaceous herbivorous sphenodontian reptile. J Paleontol 1981;55:586–597.

12 Apesteguía S, Novas FE: Large Cretaceous sphenodontian from Patagonia provides insight into lepidosaur evolution in Gondwana. Nature 2003;425:609–612.

13 Robinson PL: How *Sphenodon* and *Uromastix* grow their teeth and use them; in Bellairs Ad'A, Cox CB (eds): Morphology and Biology of the reptiles. London, Academic Press, 1976, pp 43–64.

14 Gorniak GC, Rosenberg HI, Gans C: Mastication in the Tuatara, *Sphenodon punctatus* (Reptilia: Rhynchocephalia): structure and activity of the motor system. J Morphol 1982;171:321–353.

15 Jones MEH: Tooth diversity and function in the Rhynchocephalia (Diapsida: Lepidosauria); in Barrett PM, Evans SE (eds): Ninth International Symposium on Mesozoic Terrestrial Ecosystems and Biota. London, Natural History Museum, 2006, pp 55–58.

16 Walls GY: Feeding ecology of the tuatara (*Sphenodon punctatus*) on Stephens Island, Cook Strait. NZ J Ecol 1981;4:89–97.

17 Evans AR, Sanson GD: The effect of tooth shape on the breakdown of insects. J Zool Lond 1998;246:391–400.

18 Heckert AB: Late Triassic microvertebrates from the lower Chinle Group (Otischalkian-Adamanian: Carnian) southwestern USA. NM Mus Nat Hist Sci Bull 2004;27:1–170.

19 Fraser NC: A new Sphenodontian from the early Mesozoic of England and North America: implications for correlating early Mesozoic continental deposits; in Lucas SG, Morales M (eds): The nonmarine Triassic. N Mex Museum Nat Hist Sci Bull 1993;3:135–139.

20 Fraser NC: New Triassic sphenodontids from South West England and a review of their classification. Palaeontology 1986;29:165–186.

21 Reynoso VH: A Middle Jurassic *Sphenodon*-like sphenodontian (Diapsida: Lepidosauria) from Huizachal Canyon, Tamaulipas, Mexico. J Vert Paleontol 1996;16:210–221.

22 Evans SE, Sigogneau-Russell D: New sphenodontians (Diapsida: Lepidosauria: Rhynchocephalia) from the early Cretaceous of North Africa. J Vert Paleontol 1997;17:45–51.

23 Lucas PW, Luke DA: Chewing it over: basic principles of food breakdown; in Chivers DK, Wood BA, Bilsborough A (eds): Food Acquisition and Processing in Primates. New York, Plenum Press, 1984, pp 283–301.

24 Evans AR, Sanson GD: The tooth of perfection: functional and spatial constraints on mammalian tooth shape. Biol J Linn Soc 2003;78:173–191.

25 Freeman PW, Lemen C: Puncturing ability of idealized canine teeth: edged and non-edged shanks. J Zool 2006;269:51–56.

Dr. Marc E.H. Jones
Research Department of Cell and Developmental Biology, University College London
Gower Street, UCL
London WCIE 6BT (UK)
Tel. +44 20 7679 7698, Fax +44 20 7679 7349, E-Mail marc.jones@ucl.ac.uk

Koppe T, Meyer G, Alt KW (eds): Comparative Dental Morphology.
Front Oral Biol. Basel, Karger, 2009, vol 13, pp 16–22

Molar Crown and Root Size Relationship in Anthropoid Primates

Kornelius Kupczik · Anthony J. Olejniczak · Matthew M. Skinner · Jean-Jacques Hublin

Department of Human Evolution, Max Planck Institute for Evolutionary Anthropology, Leipzig, Germany

Abstract

Mandibular corpus form is thought to reflect masticatory function and the size of the dentition, but there is no universal association between crown dimensions and corpus size across anthropoids. Previous research was based on the assumption that crown size is an appropriate proxy for overall tooth size, but this hypothesis remains largely untested. This study assesses the relationship between the volume and surface area of molar crowns and roots by examining two main hypotheses: (1) crown size correlates significantly with root size, and (2) the proportion of root-to-crown surface area is related to dietary proclivity. Permanent M_2s (n = 58) representing 19 anthropoid species were CT scanned and the volume and surface area of the crown and root were measured. Interspecific correlation and regression analyses reveal significant isometric relationships between crown and root volume and a positive allometric relationship between root and crown surface area (i.e. as crown surface area increases, root surface area becomes disproportionately greater). Intraspecifically, crown and root surface area correlate significantly in some species where such analyses were possible. In general, hard object feeders exhibit relatively larger root surface area per unit crown surface area compared to soft and tough object feeders. The results also show that despite differences in food specialization closely related species have similar root-to-crown surface area proportions, thus indicating a strong phylogenetic influence. Since it is possible that, at least in some species, crown and root size vary independently, future studies should elucidate the relationship between tooth root size and mandible form.

Primate mandibular corpus form is thought to reflect both masticatory function and the size of the developing and fully formed deciduous and permanent dentitions. A recent study, however, showed no universal association between tooth size and mandibular corpus size across anthropoids [1]. This study was based on the assumption that crown size is an appropriate proxy for overall tooth size and that crown and root size are correlated with one another, but these hypotheses remain largely untested in primate species other than *Homo sapiens*. Studies of modern human teeth have indicated positive but low correlations between linear dimensions of crowns and roots [2, 3]. Spencer [4], however, concluded from a study on tooth root form in platyrrhine seed-eaters that changes in root size may be independent of crown size changes.

The tooth roots are anchored to the alveolar bone by collagenous Sharpey's fibres. They are crucial in providing support for the tooth when it is subjected to occlusal forces. Variations in tooth root size, and in particular root surface area (SA), are adaptively linked to differences in food consistency in both platyrrhines and hominids, including the fossil giant ape, *Gigantopithecus blacki* [4, 5]. Kovacs [6] argued that a transmission of occlusal forces occurs when occlusal and root areas are equal. In contrast, forces are distributed over the root surface when the SA of the roots exceeds that of the crown. Hence, it could be argued that the latter scaling relationship would be adaptively advantageous when resistant foods are consumed which require relatively larger occlusal forces than less resistant foods. Hard object feeding specialists, such as *Cebus apella* or *Lophocebus albigena* which are known to process stiff, thick-shelled seeds with their postcanines [7–9], are therefore expected to have relatively large molar root SA for a given crown SA. In contrast, soft object feeders (e.g. *Hylobates muelleri* [10]) will have relatively less root SA, since lower bite forces are required to break down soft foods such as mature fruits. Tough object feeders like the colobines [11], which consume leaves and seeds with thin, flexible (i.e. not stiff) coats, are expected to fall between hard and soft object feeders in terms of root-to-crown SA ratio.

We aim to assess the scaling relationship between crown and root size in a range of anthropoids of varying dietary proclivity. We investigate both the volume and surface area of crowns and roots as these variables adequately reflect both the overall size of a tooth and its dietary function. We examine two main hypotheses: (1) molar crown size correlates significantly with root size, and (2) the proportion of molar root-to-crown SA is related to dietary proclivity.

Materials and Methods

The mixed-sex sample consisted of 58 relatively unworn permanent mandibular second molars representing 18 extant (Ceboidea, Cercopithecoidea, Hominoidea) and one fossil (*Gigantopithecus blacki*, Hominoidea) anthropoid species (table 1). Taxa were grouped into three dietary categories (hard, soft and tough) according to what foods are typically processed with the postcanines.

The ceboid and cercopithecoid molars (n = 47) were scanned with a Skyscan micro-CT machine yielding image stacks with a resolution of between 7 and 14 μm^3. The hylobatid molars were scanned with a Scanco micro-CT 20 system with a voxel size of between 6 and 11 μm^3. Each dataset was processed in Amira 4.1.2 (Mercury Computer Systems) and resampled to a voxel size of between 10 and 30 μm^3. Following filtering of the image stack with a combined median and mean of least variance filter, the enamel, dentine and pulp were segmented using a semiautomatic threshold-based approach. The sample also included the 3D visualizations of two *Papio anubis* and 21 hominid *(P. troglodytes, G. gorilla, P. pygmaeus, G. blacki* and *H. sapiens)* M_2s based on medical CT scans taken on a Siemens Somatom Plus 4 with a voxel size of between 0.15 mm and 0.38 mm^3 (table 1). Eleven of the hominid molars were already used in a previous study [5]. The quantification of crown and root SA (mm^2) was carried out by rendering an unsmoothed triangulated surface model. Each segmented molar was then divided into its anatomical crown and root parts by using a best-fit plane to virtually cut through the molar at the cemento-enamel junction [5] and the volumes of each part (in mm^3) were computed. The area of the cut plane between crown and root served as a proxy for the occlusal area (mm^2). The difference between the full crown area and the occlusal area is the lateral enamel area reflecting crown height. All dimensions were natural log transformed. Spearman's rank correlation coefficients (r_{rank}) and reduced major axis (RMA) regression were used to investigate bivariate trends between the parameters. Analyses were carried out in PAST v1.85.

Results

Figure 1 shows bivariate plots of M_2 root and crown volume and root and crown SA for all anthropoids. These reveal statistically nonparametric significant correlations for both variables (volume: $r_{rank} = 0.96$, $p < 0.05$; SA: $r_{rank} = 0.97$, $p < 0.05$). In figure 1a, the RMA slope is not significantly greater than 1 ($p > 0.05$), indicating an isometric scaling relationship between root and crown volume. In contrast, the RMA slope for SA is significantly greater than 1 ($p < 0.05$; fig. 1b), i.e. root SA scales with positive allometry to

Table 1. Anthropoid taxa included in study

Taxon		Dietary category	N (M_2)	Scanner
Ceboidea				
	Cebus apella	H	3	μCT
	Chiropotes satanas	S/T	1	μCT
Cercopithecoidea				
	Cercocebus torquatus	H	1	μCT
	Cercopithecus mitis	T	4	μCT
	Lophocebus albigena	H	2	μCT
	M. fascicularis	S	6	μCT
	M. sylvanus	T	1	μCT
	Papio anubis	H	5	μCT
	Colobus guereza	T	2	μCT
	Presbytis melalophos	T	3	μCT
	Trachypithecus cristata	T	1	μCT
	T. vetulus	T	2	μCT
Hominoidea				
	Hylobates muelleri	S	2	μCT
	Symphalangus syndactylus	T/S	2	μCT
	Pongo pygmaeus	H/T	2	medCT
	Gorilla gorilla	T	2	medCT
	Pan troglodytes	S	4	medCT
	Gigantopithecus blacki	H	1	medCT
	Homo sapiens	S	14	medCT (n=12) and μCT (n=2)

H = Hard object feeder; S = soft object feeder; T = tough object feeder; μCT = micro-CT; medCT = medical CT.

crown SA. Of the three largest single-taxon subsamples, *H. sapiens* (n = 14), *M. fascicularis* (n = 6) and *P. anubis* (n = 5), there are significant correlations between crown and root area in *M. fascicularis* ($r_{rank} = 0.86$, $p < 0.05$) and *P. anubis* ($r_{rank} = 0.9$, $p < 0.05$), but none for crown-root volumes. The small sample sizes of *M. fascicularis* and *P. anubis* do not facilitate testing whether RMA slopes are significantly different from a slope of 1.

Figure 2 presents the relative proportions of the occlusal area, lateral enamel area and root SA in ceboids, cercopithecoids and hominoids. The sum of the occlusal and the lateral enamel area is the full crown SA. Across the anthropoids, root SA makes up between 51 and 71% of total tooth SA. Among the cercopithecoids, the colobines exhibit relatively small root SAs (51 to 56%) compared to the cercopithecines (58 to 68%). Relative root SA proportions of the two ceboid taxa *(C.*

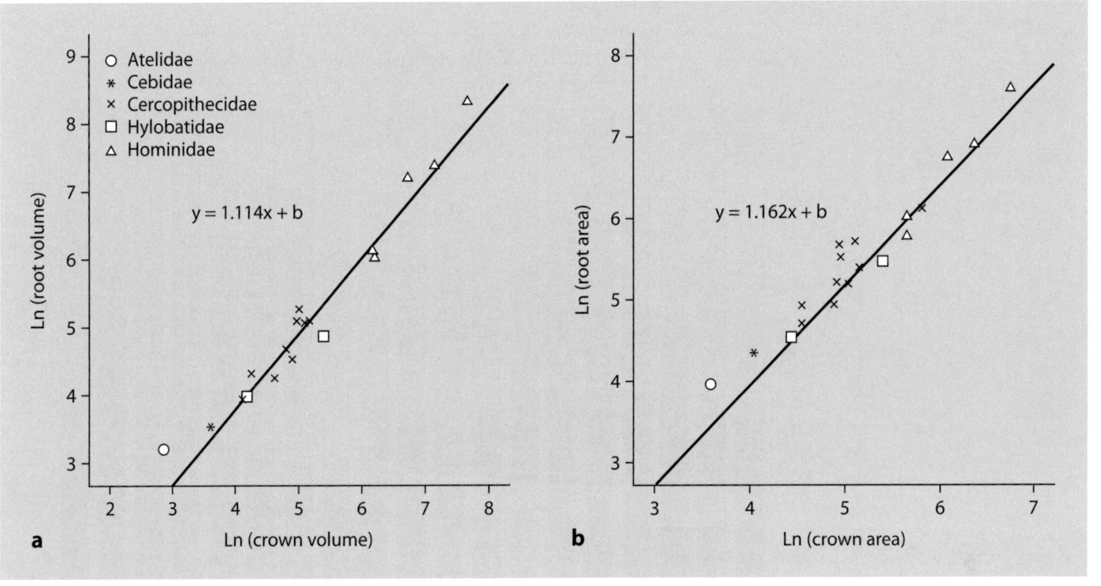

Fig. 1. Bivariate plots of M_2 root against crown volume (left) and M_2 root against crown surface area (right). Values are species-means. Note that crown and root volumes scale isometrically (RMA slope is not significantly different from 1), while the root surface area scales with positive isometry with crown surface area (RMA slope is significantly larger than 1).

satanas and *C. apella)* fall between the two cercopithecoid subfamilies (59 and 57%, respectively). Among hominoids, both hylobatids and modern humans have markedly less relative root SA (53 and 55%, respectively) than the extant great apes (60 to 67%) and the extinct *G. blacki* (70%). Hylobatids and humans also show the relatively smallest occlusal area of all taxa (between 23 and 27% of total crown SA).

When the dietary categories are taken into account, hard object feeders have the largest root SA relative to a given crown SA within and across taxonomic groups (e.g. *C. torquatus, L. albigena, P. pygmaeus* and *G. blacki*). With the exception of *C. apella* and *P. anubis*, hard object feeding taxa have between 64 and 70% root SA. In contrast, within each of the subfamilies soft object feeders have relatively small roots with SAs of less than 60% (e.g. *H. muelleri* among hominoids and *M. fascicularis* among cercopithecines). Tough object feeders range in relative root SA between 51% *(T. vetulus)* and 65% *(M. sylvanus)*, which is less than some of the soft object feeders.

Discussion

The results support, to some extent, the hypothesis that crown and root size are correlated with one another (cf. [1]). Both volumes and SAs of crowns and roots show a strong positive correlation across primates. In cases where sample size was sufficiently large, intraspecific analysis revealed significant correlations between root and crown SAs in two of the three species *(M. fascicularis* and *P. anubis* but not in *H. sapiens).* The latter is strengthened by the finding that root and crown SA scale with positive allometry, i.e. changes in root area are disproportionally larger than changes in crown area (fig. 1). An increase

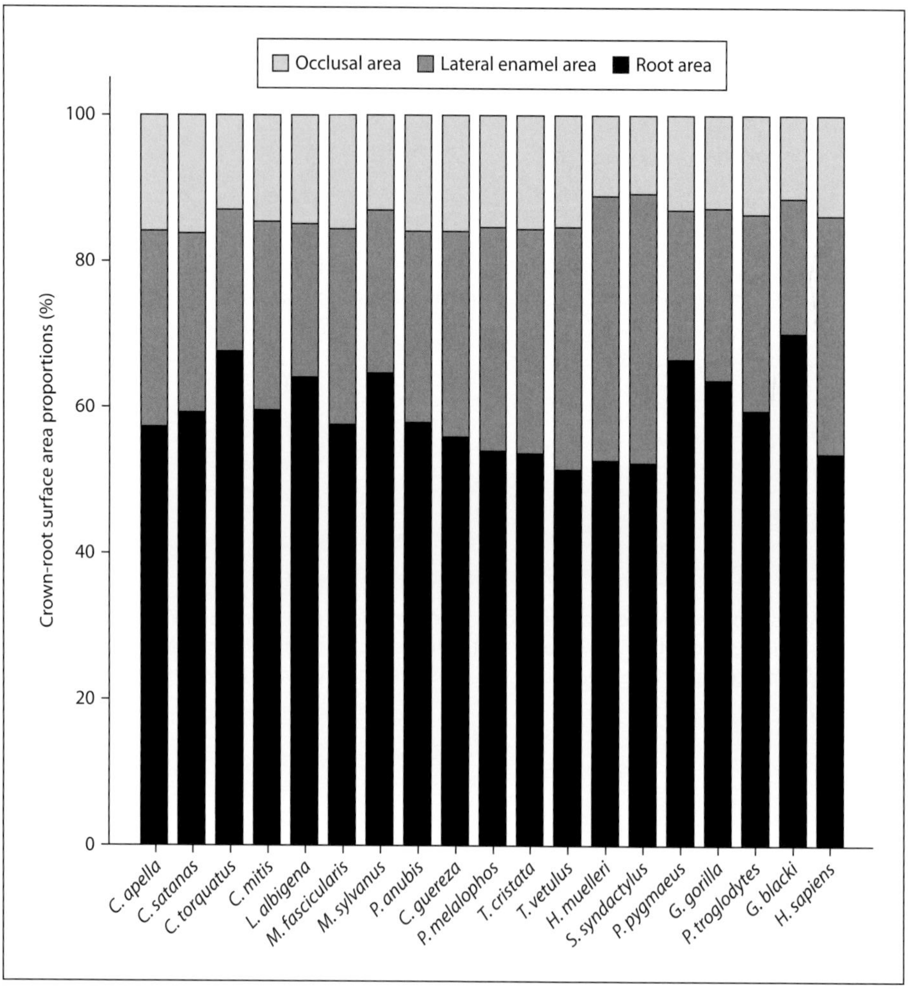

Fig. 2. Relative proportions of root surface area, occlusal area and lateral enamel area in anthropoids.

in root SA is likely to be of adaptive advantage because it allows the comminution of mechanically more resistant foods, as has been suggested previously [4, 5] and is bolstered by the present data. In support of the second hypothesis, among catarrhines a high root-to-crown area ratio is generally favoured by those taxa engaging in hard object feeding, whereas soft object feeders have a low ratio (fig. 2). The relatively low root SAs of the hard object feeding *C. apella* compared to other hard object feeders and to *C. satanas*, which uses its anterior dentition rather than its molars to open seeds [7, 12], is at odds with our expectations, but confirms similar results reported by Spencer [4]. The results for tough object feeders, in particular the colobines, are only partially in agreement with our expectations. It may be that colobines exert relatively low occlusal forces to fracture leaves with their wedge-shaped molar crowns [11], which in turn would require less tooth support. Among extant great apes, *P. pygmaeus* exhibits larger relative root SA than the African great apes. This concurs with recent studies showing that *P. pygmaeus* consumes, on average, tougher and stiffer foods than *G. beringei* and *P. troglodytes* [13, 14]. In particular, Taylor et al. [13] suggest that the maximum toughness of non-fruit, non-leaf vegetation may be the critical mechanical property affecting the load resistance abilities of the mandible and perhaps, by extension, the teeth.

The results also reveal that despite differences in food specialisation closely related species (e.g. within the ceboids, colobines and hylobatids) have similar root-to-crown SA proportions. This indicates that phylogeny is likely to have a strong influence on crown/root size.

It also cannot be excluded that, at least in some species, crown and root size vary independently as has been suggested by Spencer [4] and confirmed here for *H. sapiens*. Tooth size, and in particular root size, may be a function of mandibular corpus form [15, 16]. There is mixed support for this hypothesis in Plavcan and Daegling's study [1] who, albeit observing significant intraspecific correlations in some species, did not find a universal association between molar crown and corpus dimensions across anthropoids. Future studies incorporating canine and molar root form should be able to elucidate this issue further.

Acknowledgements

We are grateful to Frieder Mayer and Henrik Turni (Museum für Naturkunde, Berlin) for access to specimens in their care. This research was supported by the Max-Planck-Gesellschaft.

References

1 Plavcan JM, Daegling DJ: Interspecific and intraspecific relationships between tooth size and jaw size in primates. J Hum Evol 2006;51:171–184.
2 Garn SM, Van Alstine WL, Cole PE: Relationship between root lengths and crown diameters of corresponding teeth. J Dent Res 1978;57:636.
3 Ozaki T, Satake T, Kanazawa E: Morphological significance of root length variability in comparison with other crown dimensions. II. Correlation between crown and root measurements. J Nih Univ S Dent 1988;30:11–21.
4 Spencer MA: Tooth-root form and function in platyrrhine seed-eaters. Am J Phys Anthrop 2003;122:325–335.
5 Kupczik K, Dean MC: Comparative observations on the tooth root morphology of *Gigantopithecus blacki*. J Hum Evol 2008;54:196–204.
6 Kovacs I: A systematic description of dental roots; in Dahlberg AA (ed): Dental Morphology and Evolution. Chicago, University of Chicago Press, 1971, pp 211–256.
7 Kinzey WG: Ceboid models for the evolution of hominoid dentition. J Hum Evol 1974;3:193–203.
8 Kay RF: The nut-crackers: a new theory of the adaptations of the Ramapithecinae. Am J Phys Anthrop 1981;55:141–151.
9 Lambert JE, Chapman CA, Wrangham RW, Conklin-Brittain NL: Hardness of cercopithecine foods: implications for the critical function of enamel thickness in exploiting fallback foods. Am J Phys Anthrop 2004;125:363–368.
10 Palombit RA: Inter- and intraspecific variation in the diets of sympatric siamang (*Hylobates syndactylus*) and lar gibbons (*Hylobates lar*). Folia Primatol 1997;68:321–337
11 Lucas PW, Teaford MF: Functional morphology of colobine teeth; in Davies AG, Oates JF (eds): Colobine Monkeys: Their Ecology, Behaviour and Evolution. Cambridge, Cambridge University Press, 1994, pp 173–203.
12 Martin LB, Olejniczak AJ, Maas MC: Enamel thickness and microstructure in pitheciin primates, with comments on dietary adaptations of the middle Miocene hominoid *Kenyapithecus*. J Hum Evol 2003;45:351–367.
13 Taylor AB, Vogel ER, Dominy NJ: Food material properties and mandibular load resistance abilities in large-bodied hominoids. J Hum Evol 2008;55:604–616.

14 Vogel ER, van Woerden JT, Lucas PW, Utami Atmoko SS, van Schaik CP, Dominy NJ: Functional ecology and evolution of hominoid molar enamel thickness: *Pan troglodytes schweinfurthii* and *Pongo pygmaeus wurmbii*. J Hum Evol 2008;55:60–74.

15 Wolpoff MH: Some aspects of human mandibular evolution; in McNamara JA Jr (ed): Determinants of Mandibular Form and Growth. Craniofacial Growth Series. Ann Arbor, Center for Human Growth and Development, University of Michigan, 1975, vol 4, pp 1–64.

16 Wood BA: Allometry and hominid studies; in Bishop WW (ed): Geological Background to Fossil Man. Edinburgh, Scottish Academic Press, 1978, pp 125–128.

Dr. Kornelius Kupczik
Max Planck Institute for Evolutionary Anthropology, Department of Human Evolution
Deutscher Platz 6
DE–04103 Leipzig (Germany)
Tel. +49 341 3550 756, Fax +49 341 3550 399, E-Mail kornelius.kupczik@eva.mpg.de

Koppe T, Meyer G, Alt KW (eds): Comparative Dental Morphology.
Front Oral Biol. Basel, Karger, 2009, vol 13, pp 23–29

How Many Landmarks? Assessing the Classification Accuracy of *Pan* Lower Molars Using a Geometric Morphometric Analysis of the Occlusal Basin as Seen at the Enamel-Dentine Junction

Matthew M. Skinner[a,b] · Philipp Gunz[a] · Bernard A. Wood[b] · Jean-Jacques Hublin[a]

[a]Department of Human Evolution, Max Planck Institute for Evolutionary Anthropology, Leipzig, Germany; [b]Center for the Advanced Study of Hominid Paleobiology, Department of Anthropology, George Washington University, Washington, DC, Wash., USA

Abstract

Previous research has demonstrated that species and subspecies of extant chimpanzees and bonobos can be distinguished on the basis of the shape of enamel-dentine junction of lower molar crowns. Thus, there is potential for fossil taxa, particularly fossil hominins, to be distinguished at similar taxonomic levels using lower molar crown morphology. New imaging techniques allow for the collection of large amounts of shape data, but it is not clear whether taxonomic distinctiveness increases with the inclusion of more and more finely detailed aspects of crown shape. We examine whether increasing the amount of shape data collected will lead to an increase in the accuracy with which enamel-dentine junction (EDJ) shape classifies *Pan* lower first and second molars at the species and subspecies level. Micro-computed tomography was employed to non-destructively image the EDJ and geometric morphometric analytical methods were used to compare EDJ shape among samples of *Pan paniscus*, *Pan troglodytes troglodytes*, and *Pan troglodytes verus*. The results of discriminant analyses using three landmark sets (number of landmarks = 8, 112, and 534 landmarks and semi-landmarks, respectively) indicate a high degree of classification accuracy for each landmark set, with small increases in accuracy as the numbers of landmarks are increased. The morphological differences in EDJ shape among the taxa are subtle, but consistent, and relate to the relative height and position of the dentine horns. Thus, EDJ shape can contribute to taxonomic analyses and the more information that can be included the better.

It has long been acknowledged that the enamel-dentine junction (EDJ), which underlies the enamel cap of primate teeth, carries information about the original shape of the outer enamel surface of the crowns or worn teeth [1–8] and that it can be used as a source of taxonomically relevant data [9–12]. In a recent analysis, it was demonstrated that lower molar EDJ shape distinguishes both species and subspecies of the genus *Pan* (chimpanzees and bonobos) with a high degree of reliability [7]. This result was consistent with previous analyses of the shape of the outer enamel surface of extant apes [13–15] and fossil hominins [e.g., 16, 17].

Whereas in past decades only linear dimensions were used to summarize tooth crown shape, the use of microCT imaging and geometric morphometrics now allows the collection of shape data at a hitherto unprecedented level of detail. The collection of such datasets can be expensive, time consuming and computationally difficult. This begs the question: how many landmarks (capturing what level of detail) are necessary to reliably distinguish closely related taxa? In this contribution we investigate whether increasing the number of landmarks collected on the EDJ surface of *Pan* lower molars will increase the accuracy with which they are classified at both the species and subspecies level. The results of this analysis can guide future analyses of fossil taxa for whom the analysis of tooth shape contributes strongly to assessments of taxonomic affiliation and phylogenetic relationships.

Methods

Two species of chimpanzee are commonly recognized and their distinction is supported by both morphological (e.g. [13–15, 18]) and molecular studies [19, 20]. *Pan paniscus*, also known as the bonobo or pygmy chimpanzee, is found in the Democratic Republic of the Congo and all but the southern limits of its range are defined by the Congo River. Several subspecies of *Pan troglodytes* are commonly recognized, and their ranges are separated by geographic barriers. For example, *Pan troglodytes verus* – the western chimpanzee – is separated from other *Pan* populations by the Dahomey gap, *Pan troglodytes vellorosus* – the Nigerian chimpanzee – by the Sanaga River, *Pan troglodytes troglodytes* – the central chimpanzee – by the Ubangi River, and *Pan troglodytes schweinfurthii* – the eastern chimpanzee – by the Ubangi and Congo Rivers. While the subspecies distinction of each of these taxa is debated [21] and is more strongly supported for some taxa (e.g. *P. t. verus*) than for others (e.g. the distinction between *P. t. troglodytes* and *P. t. schweinfurthii*), both morphological [14, 15, 18] and molecular evidence [20] has been used to support their distinction.

The sample includes lower first and second molars of *P. paniscus* (Pp; n = 17) and two subspecies of *P. troglodytes* (Pt): *P. t. troglodytes* (Ptt; n = 15) and *P. t. verus* (Ptv; n = 16). The Ptt sample derives from the Museum für Naturkunde in Berlin (ZMB), Germany, the Ptv sample derives from a skeletal collection housed at the Max Planck Institute for Evolutionary Anthropology in Leipzig, Germany, and the Pp sample derives from the Royal Museum for Central Africa in Tervuren, Belgium (MRAC). Taxonomic affiliation is based on locality information and museum catalogue information associated with each specimen. A more detailed description of the study sample and analytical methods can be found in previous publications [7, 12].

Each tooth was microCT scanned using a SKYSCAN 1172 Desktop Scanner and raw projections were converted into TIFF image stacks using NRecon. Pixel dimensions and slice spacing of the reconstructed TIFF image stacks ranged between 10 and 20 μm. To facilitate tissue segmentation, the complete image stack for each molar was filtered using a three-dimensional median filter (kernel size of 3) followed by a mean of least variance filter (kernel size of 3). After segmentation, the EDJ was reconstructed as a triangle-based surface model. Each EDJ surface was oriented manually into its anatomical position. The occlusal surface of the EDJ was isolated by creating a section plane parallel to the occlusal surface and removing the sides of the EDJ from a plane below the lowest point in the occlusal basin.

Collection and Processing of Surface Landmarks

We analyzed three sets of landmarks (fig. 1). The first set (referred to as 'MAIN') included eight landmarks: one on the tip of the dentine horn of each primary cusp (i.e. protoconid, metaconid, entoconid, hypoconid, and hypoconulid), one at the mid-point on the marginal crest connecting the protoconid and metaconid, and one on the lowest point on the marginal ridges between the protoconid and hypoconid, and the hypoconid and hypoconulid, respectively. The second set (referred to as 'RIDGE') includes coordinates (approximately 50–70) along the tops of the ridges that connect the five dentine horns. This set of points forms a continuous line, beginning at the tip of the protoconid and moving in a lingual direction. The third set (referred to as 'occlusal basin') includes coordinates located within the occlusal basin of the EDJ border by the marginal ridge. For the latter two analyses only the four dentine horn tips were treated as landmarks, all other points were treated as semi-landmarks [22].

The coordinates collected on the ridge and occlusal basin landmark sets were treated as semi-landmarks on curves and surfaces, respectively. We used the algorithm described by Gunz et al. [23] that allows semi-landmarks to slide along tangents to ridge curves and tangent planes to the surface. Semi-landmarks were iteratively slid to minimize the bending energy of the thin-plate spline interpolation function computed between each specimen and the sample Procrustes average. After each sliding step the semi-landmarks were projected back onto the original surface. After convergence of the sliding algorithm, these semi-landmarks were considered homologous for the purpose of multivariate analyses.

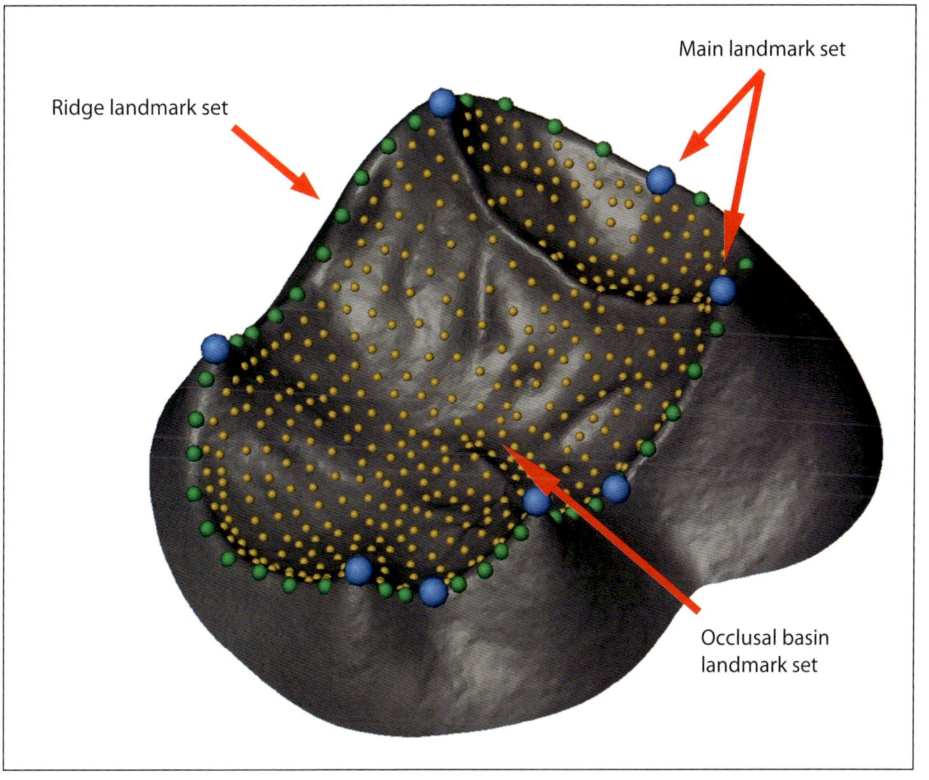

Fig. 1. EDJ surface model of a lower molar illustrating the three landmarks sets used to capture EDJ shape. *Main* landmarks are collected on the tips of the dentine horns and in the troughs between the mesial and buccal dentine horns (large blue spheres). The *ridge* landmark set included an arbitrary number of points collected along the ridge that runs between the dentine horns (small green spheres). The *occlusal basin* landmark set included the surface within the confines of the marginal ridge (small yellow spheres). Points illustrated here are representative of those collected on the original specimens and are not the same as the interpolated semi-landmarks (see text for details).

The landmarks and semi-landmarks were then converted to shape coordinates by generalized least squares Procrustes superimposition. This removed information about location and orientation from the raw coordinates and standardized each specimen to unit centroid size, a size-measure computed as the square root of the sum of squared Euclidean distances from each (semi-)landmark to the specimen's centroid. All data preprocessing was done in Mathematica v6.0 (www.wolfram.com) using software written by PG.

Principal component analysis (PCA) of shape coordinates [24] was used to examine overall shape variation in the sample and the distribution of each group in shape space. Canonical variate analysis (CVA), computed in a subspace of the first principal components (8–12), was used to assess the accuracy with which molars were correctly classified to taxon [for detailed discussion of this issue, see 26]. We used a cross-validation approach in which each specimen was considered unknown and then classified based on the remaining sample. The PCA, CVA, and classifications were implemented in the software package R and groups were assigned equal prior probabilities.

To visualize EDJ shape variation between taxa, we employed a method that allows a 3D triangulated surface reconstruction of the EDJ to be deformed to match the mean molar configuration of each taxon [12, 23, 25]. First, the EDJ surface model of one specimen was chosen at random. We then warped the vertices of this surface

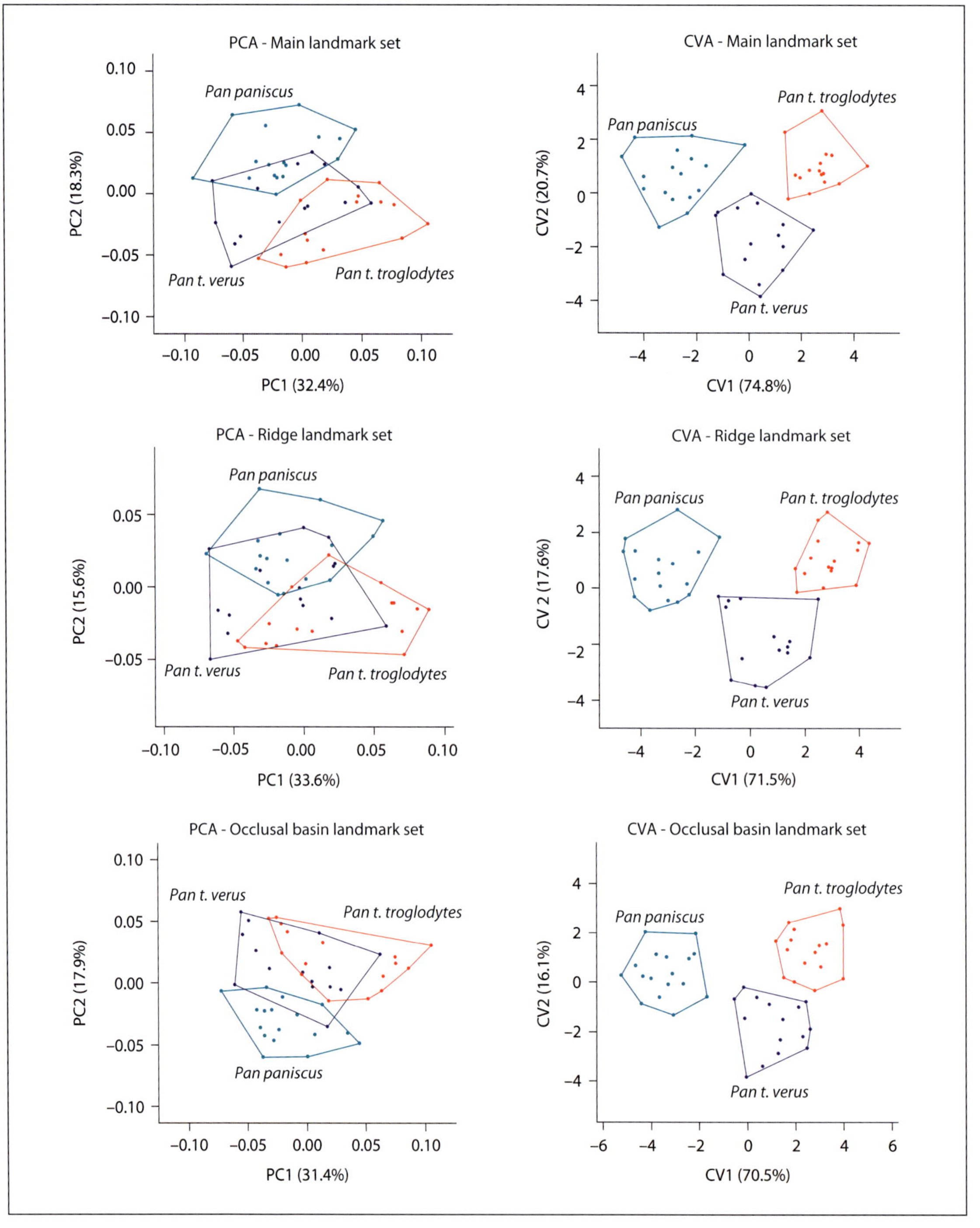

Fig. 2. Plots of the principal component analyses (PCA) and canonical variates analyses (CVA) performed on each landmark set. The percentage of total shape variation is listed in brackets for each PC or CV axis, respectively. Ten PCs were used for each CVA plot presented here although classification accuracy was assessed using each of 8–12 PCs.

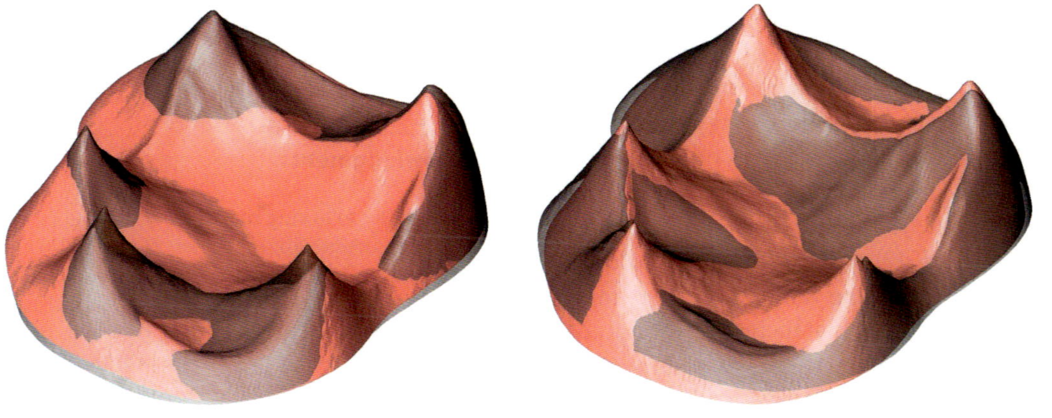

Fig. 3. Taxonomic differences in mean EDJ shape. Left: species level comparison between the mean *Pan paniscus* molar shape (transparent) and the mean shape of the combined *Pan troglodytes* molar sample (solid); Right: subspecies level comparison between the mean *Pan t. verus* molar shape (transparent) and the mean *Pan t. troglodytes* molar shape (solid). Note subtle differences in dentine horn height and positioning on the EDJ.

into Procrustes space using the thin-plate spline interpolation function between the landmark configuration of this specimen and the Procrustes average configuration of the whole sample. Finally, we computed a thin-plate spline between this mean configuration and each target form (e.g. the mean configuration of the Ptv M_1 sample) to produce a surface model of the appropriate mean shape. In order to visualize the taxonomic differences at each molar position the mean shapes were superimposed in the software package Amira with one surface rendered transparent for better visual comparison.

Results

The PCA and CVA of Procrustes shape coordinates of the molar sample for each landmark set are illustrated in figure 2; the percentages of total shape variation explained by each PC and CV are listed in parentheses. We find overlap of the groups in the PCA but complete separation along the first two CV axes. This indicates that there are consistent, but small-scale, differences in shape between the taxa. The spatial patterning in the PCA and CVA for each landmark set is very similar. Using a cross-validation analysis of the CV scores the accuracy of classification to species for each landmark set is as follows: main = 88–94%; ridge = 96%; occlusal basin = 92–100%. The accuracy of classification to subspecies is: main = 83–90%; ridge = 85–92%; occlusal basin = 88–96%. The mean EDJ molar shape of *Pan paniscus* compared that of the combined *Pan troglodytes* sample is visualized in figure 3 (Left). This represents the shape differences at the species level and includes relatively tall distal dentine horns (entoconid and hypoconulid) and a relatively deep occlusal basin in *Pan paniscus* compared to *Pan troglodytes*. The mean shape differences in the two subspecies, *Pan t. troglodytes* and *Pan t. verus*, are also illustrated in figure 3 (right). As might be expected between subspecies the mean molar EDJ shapes are quite similar with only minor variation in the relative position and height of the dentine horns.

Conclusions

This study investigated whether increasing the number of 3D coordinates used to capture the shape of the EDJ of lower molars increases the accuracy with which molars are correctly classified to species and subspecies of *Pan*. These results indicate (1) that classification accuracy is higher at the species level compared to the subspecies level, (2) that classification accuracy is relatively high using only a limited number of landmarks, and (3) that increasing the number of landmarks collected on the EDJ results in slightly higher classification accuracy at both the species and subspecies level. Our results suggest that EDJ morphology carries taxonomically relevant information even if only a limited number of landmarks are available for data collection, and that because they are the sister clade to the panins the same conclusions are likely to apply to the hominin clade.

Acknowledgments

We thank the following museums and curators for access to specimens: Robert Asher, Hendrik Turni and Irene Mann of the Museum für Naturkunde, Berlin, Germany; Emmanuel Gilissen and Wim Wendelen of the Royal Museum for Central Africa, Tervuren, Belgium; Christophe Boesch, MPI-EVA, Leipzig, Germany. Hella Issler, Heiko Temming, Gert Wollny, and Philipp Mitteroecker are also thanked for their assistance. This research was supported by NSF IGERT, EVAN Marie Curie Research Training Network MRTN-CT-019564, and the Max Planck Society.

References

1 Korenhof CAW: Morphogenetical Aspects of the Human Upper Molar. Utrecht, Uitgeversmaatschappij Neerlandia, 1960.

2 Nager G: Der Vergleich zwischen dem räumlichen verhalten des Dentin-kronenreliefs und dem Schmelzrelief der Zahnkrone. Acta Anat 1960;42:226–250.

3 Kraus BS, Jordan R: The Human Dentition before Birth. Philadelphia, Lea & Febiger, 1965.

4 Sakai T, Hanamura H: A morphology study of enamel-dentin border on the Japanese dentition. VII. General conclusion. J Anthropol Soc Nippon 1973;81:87–102.

5 Corruccini RS: The dentinoenamel junction in primates. Int J Primatol 1987;8:99–114.

6 Schwartz GT, Thackeray JF, Reid C, van Reenen JF: Enamel thickness and the topography of the enamel-dentine junction in South African Plio-Pleistocene hominids with special reference to the Carabelli trait. J Hum Evol 1998;35:523–542.

7 Skinner M: Enamel-dentine junction morphology in extant hominoids and fossil hominins; PhD Diss, George Washington University, 2008.

8 Skinner MM, Wood BA, Boesch C, Olejniczak AJ, Rosas A, Smith TM, Hublin J-J: Dental trait expression at the enamel-dentine junction of lower molars in extant and fossil hominoids. J Hum Evol 2008;54:173–186.

9 Corruccini RS: The dentino-enamel junction in primate mandibular molars; in Lukacs JR (ed): Human Dental Development, Morphology, and Pathology: A Tribute to Albert A Dahlberg. Portland, University of Oregon Anthropological Papers, 1998, vol 54, pp 1–16.

10 Olejniczak AJ, Gilbert CC, Martin LB, Smith TM, Ulhaas L, Grine FE: Morphology of the enamel-dentine junction in sections of anthropoid primate maxillary molars. J Hum Evol 2007;53:292–301.

11 Macchiarelli R, Bondioli L, Debénath A, Mazurier A, Tournepiche J-F, Birch W, Dean C: How Neanderthal molar teeth grew. Nature 2006;444:748–751.

12 Skinner MM, Gunz P, Wood BA, Hublin J-J: Enamel-dentine junction (EDJ) morphology distinguishes the lower molar molars of *Australopithecus africanus* and *Paranthropus robustus*. J Hum Evol 2008;55:979–988.

13 Johanson DC: An odontological study of the chimpanzee with some implications for hominoid evolution; PhD Diss, University of Chicago, 1974.

14 Uchida A: Craniodental Variation among the Great Apes. Cambridge, Peabody Museum of Archaeology and Ethnology, Harvard University, 1996.

15 Pilbrow V: Population systematics of chimpanzees using molar morphometrics. J Hum Evol 2006;51:646–662.

16 Wood BA, Abbott SA. Analysis of the dental morphology of Plio-Pleistocene hominids. I. Mandibular molars: crown area measurements and morphological traits. J Anat 1983;136:197–219.

17 Suwa G, Wood BA, White TD: Further analysis of mandibular molar crown and cusp areas in Pliocene and Early Pleistocene hominids. Am J Phys Anthropol 1994;93:407–426.

18 Shea BT, Leigh SR, Groves CP: Multivariate craniometric variation in chimpanzees: implications for species identification in paleoanthropology; in Kimbel WH, Martin LB (eds): Species, Species Concepts and Primate Evolution. New York, Plenum Press, 1993, pp 265–296.

19 Ruvolo M, Pan D, Zehr S, Goldberg T, Disotell TR, von Dornum M: Gene trees and hominoid phylogeny. Proc Natl Acad Sci USA 1994;91:8900–8904.

20 Becquet C, Patterson N, Stone AC, Przeworski M, Reich D: Genetic structure of chimpanzee populations. PLOS Genet 2007;DOI: 10.1371/journal.pgen.0030066.

21 Fischer A, Pollack J, Thalmann O, Nickel B, Pääbo S: Demographic history and genetic differentiation in apes. Curr Biol 2006;16:1133–1138.

22 Bookstein F: Landmark methods for forms without landmarks: morphometrics of group differences in outline shape. Med Image Anal 1997;1:225–243.

23 Gunz P, Mitteroecker P, Bookstein FL: Semilandmarks in three dimensions; in Slice DE (ed): Modern Morphometrics in Physical Anthropology. New York, Kluwer Academic/Plenum Publishers, 2005, pp 73–98.

24 Bookstein F: Morphometric Tools for Landmark Data: Geometry and Biology. Cambridge, Cambridge University Press, 1991.

25 Gunz P, Harvati K: The Neanderthal 'chignon': variation, integration, and homology. J Hum Evol 2007;52:262–274.

26 Skinner M, Gunz P, Wood B, Boesch C, Hublin JJ: Discrimination of extant Pan species and the subspecies using the enamel-dentine junction morphology of lower molars. Am J Phys Anthropol 2009; E-Pub ahead of print.

Dr. Matthew M. Skinner
Department of Human Evolution, Max Planck Institute for Evolutionary Anthropology
Deutscher Platz 6
DE–04103 Leipzig (Germany)
Tel. +49 341 355 0375, Fax +49 341 355 0399, E-Mail skinner@eva.mpg.de

Koppe T, Meyer G, Alt KW (eds): Comparative Dental Morphology.
Front Oral Biol. Basel, Karger, 2009, vol 13, pp 30–35

Preserved Microstructure and Mineral Distribution in Tooth and Periodontal Tissues in Early Fossil Hominin Material from Koobi Fora, Kenya

R. Furseth Klinge[a] · M.C. Dean[b*] · S. Risnes[a] · M. Erambert[c] · A.E. Gunnæs[d]

Departments of [a]Oral Biology, [c]Geosciences and [d]Physics, University of Oslo, Oslo, Norway; [b]Department of Cell and Developmental Biology, UCL, London, UK

Abstract

The aim of this study was to explore further the preservation of tissues and the mineral distribution in 1.6 million-year-old fossil hominin material from Koobi Fora, Kenya attributed to *Paranthropus boisei* (KNM-ER 1817). Bone, dentine and cementum microstructure were well preserved. Electron microprobe analysis of dentine and bone revealed an F-bearing apatite. Calcite now filled the original soft tissue spaces. The average Ca/P atomic ratio was 1.93, as compared to 1.67 in biological hydroxyapatite, indicating that the Ca-content had increased during fossilization. Analytical sums for mineral content were ~90 wt%. Some of the remaining 10wt% may be preserved organic material. Demineralized dentine fragments showed irregularly distributed tubules encircled with a fibrous-like electron-dense material. A similar material was observed in demineralized dentine. Within this, structures resembling bacteria were seen. In demineralized bone an electron-dense material with a fibrous appearance and a banding pattern that repeated every 64 nm, similar to that of collagen, was noted. SEM of an enamel fragment (KNM-ER 6081) showed signs of demineralization/remineralization. Retzius lines, Hunter-Schreger bands and prism cross-striations spaced 3.7–7.1.µm apart were noted. Prisms were arranged in a pattern 3 configuration and deeper areas containing aprismatic enamel were occasionally observed. We conclude that a great deal of informative microstructure and ultrastructure remains preserved in this fossil material. We also hypothesize that the high mineral content of the tissues may 'protect' parts of the organic matrix from degradation, since our findings indicate that some organic matrix may still be present.

Previously, we reported on the mineral content and density of bone, dentine and cementum in the fossilized remains of a poorly preserved hemi-mandibular fragment [1]. This ~1.6 Ma early hominin, attributed to *Paranthropus boisei*, was recovered from the site of Koobi Fora, Kenya in 1973 [2]. Good histological details of typical bone, dentine and cementum were previously observed with polarized light microscopy of ground sections. Energy dispersive X-ray analysis indicated that both calcium and phosphorus levels in the fossilized hard tissues were compatible with biological hydroxyapatite. The aims of this second study were several: We wanted to explore the possibility that some organic material might still remain in the fossilized tissue. A second aim was to study further the mineral composition with electron microprobe analysis, which is a more accurate method than energy dispersive X-ray analysis. A further aim was to investigate the nature and microstructure of a small fragment of enamel.

Materials and Methods

An early fossil hominid mandible (KMN-ER 1817) containing six root apices from four permanent teeth was examined. An isolated enamel fragment associated with KNM-ER 6081 approximately 1 × 2 mm was also studied. This material has been described previously [1–3] and was examined in more detail here. Ground sections were made and small blocks of tissue were dissected out and decalcified and fixed simultaneously in a solution of EDTA and paraformaldehyde. Small tissue blocks were embedded in EPON and sectioned for transmission electron microscopy (TEM). Some sections were stained with uranyl acetate and lead citrate while some sections were examined unstained in the TEM. The tissues were also studied with microradiography (MR), electron microprobe analysis (EMP) and selected area diffraction in the TEM. The enamel fragment was ground parallel with the prism direction, acid-etched and examined in the scanning electron microscope (SEM).

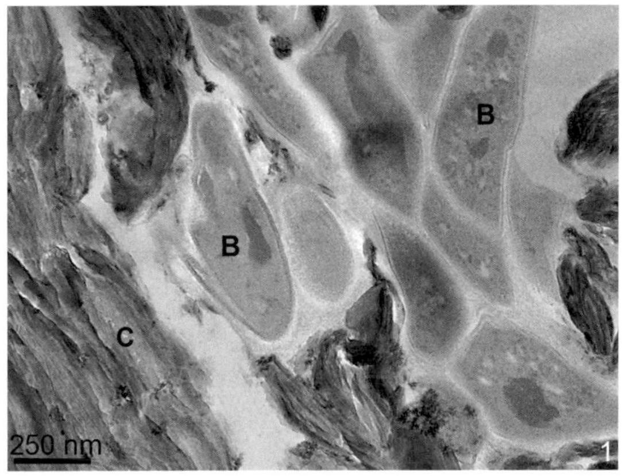

Fig. 1. Electron micrograph of EDTA demineralized dentine from the exposed aspect of the longitudinal ground section. When examined in reflected light, the area where this specimen was taken from had a brownish color and seemed degraded. An electron-dense, fibrous material (C) persisted and next to the fibrous material structures consistent in size and morphology with bacteria (B) can be seen.

Results

Microradiography of bone revealed a clear lamellar structure with numerous lacunae. The original soft tissue spaces (bone marrow, periodontal ligament and pulp) had a similar radiodensity but a lower mineral content than dentine, cementum and bone. The peripheral dentine appeared homogeneous, while dentine closer to the pulp contained numerous tubules with a straight course. Electron microprobe analysis of dentine and bone revealed that F-bearing apatite was present, but that calcite had formed in the original soft tissues spaces. The average F-content of dentine was 4.5 wt%, while that of bone was 2.8 wt%. The average Ca/P atomic ratio was 1.93 both in dentine and bone. Average analytical total mineral in dentine was 92.5 wt% and in bone 90.2 wt%.

Cutting ultrathin sections of EDTA-demineralized dentine was fairly easy. This was in contrast to mineralized, fossilized dentine which was almost impossible to section, even with diamond knives, so evidently the demineralization process had been successful. When examined in the TEM cross-sectioned dentinal tubules with an irregular distribution were observed in the pulpal part of the dentine. The diameter of the tubules varied, but was mostly 1.5–2.5 μm. Between the tubules a structureless material was observed. Often a thin zone of fibrous electron-dense material surrounded the dentinal tubules. In EDTA-demineralized, degraded dentine from the exposed outer aspect of the root, a similar electron dense material was seen. Next to this, structures whose size and morphology resembled bacteria were noted (fig. 1). Selected area electron diffraction from the fibrous material showed sharp rings indicating that it is polycrystalline. The presence of possibly more than one phase made the diffraction analysis difficult and so far we have not yet been able to identify which phases are present in the demineralized sample.

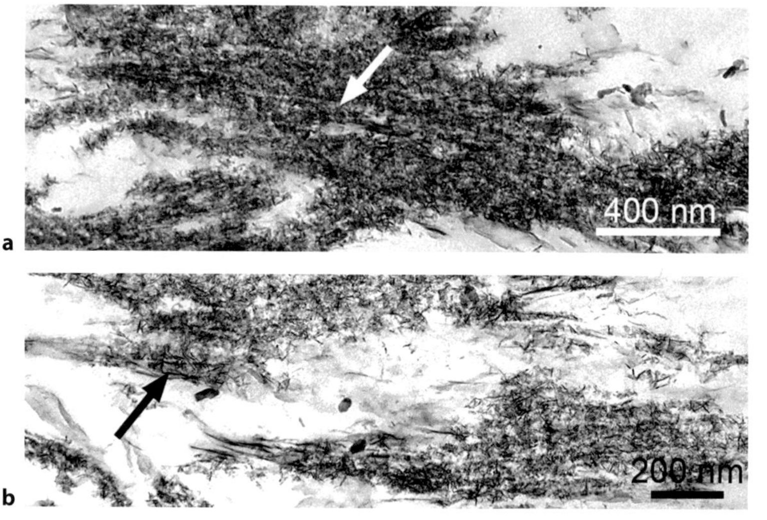

Fig. 2 a, b. Electron micrographs of EDTA demineralized alveolar bone showing an electron-dense material with a fibrous appearance from two areas in the same specimen. Within the fibrous material cross-banding typical of collagen that repeats every 64 nm can be discerned (white and black arrows, respectively).

Electron micrographs of the EDTA demineralized alveolar bone showed an electron dense material with a fibrous appearance in some areas. The areas with a fibrous appearance showed a banding pattern typical of collagen that repeated every 64 nm (fig. 2a, b).

A distinctive feature of the enamel fragment was the presence in the subsurface enamel (at about 100–200 μm) of a continuous 50–100 μm thick zone of porous enamel and just inside it a thicker zone of more dense enamel. Enamel prisms were visible on all four aspects of the ground enamel fragment. Where prisms were cut transversely, they were organized in a pattern 3 configuration. Accentuated Retzius lines were observed on all aspects. Hunter-Schreger bands were visible on one aspect of the block, which allowed the correct orientation of the fragment to be determined. Prism cross-striation spacings ranged between 3.7 and 7.1 μm with a mean value of 5.1 μm between adjacent cross-striations. Isolated regions of aprismatic enamel were occasionally observed deep to the porous/dense zones.

Discussion

This study shows that the fossil material now contains an F-bearing apatite, with 4.5 wt% F-content in dentine and 2.8 wt% in bone, although these values may be slightly overestimated [4]. An exchange to fluorapatite via fluoride ions in ground water post-mortem is quite common in fossilized bone and teeth [5] and gives a mineral phase that is more resistant to dissolution. An average Ca/P atomic ratio of 1.93 in both dentine and bone as compared to approximately 1.67 in biological apatite, shows that the Ca content in particular has increased during the fossilization process. A low analytical total of 92.5 wt% in dentine and 90.5 wt% in bone may indicate that at least some of the remaining 7.5–9.5 wt% may be organic material. The radiodensity of the various hard tissues is in accordance with our previous findings [1] and the finding of straight dentinal tubules in root dentine is similar to that observed in recent human dentine. The nature of the polycrystalline material found in decayed dentine as well as surrounding the

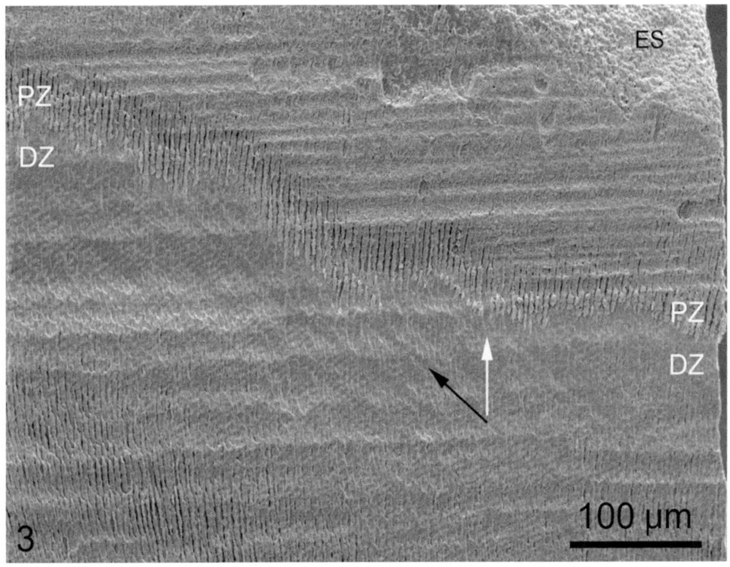

Fig. 3. SEM of the ground, then acid-etched enamel fragment. A porous enamel zone (PZ) and a dense enamel zone (DZ) at some distance from enamel surface (ES) are visible. The plane of section is probably longitudinal judged from the general direction of the Retzius lines (black arrow) relative to the enamel surface, as well as that of prism direction (white arrow). Horizontal grooves stem from preparation of the fragment surface.

dentinal tubules so far remains a puzzle, but it clearly shows that some matter, perhaps organic, survives the demineralization process. Further studies are needed to establish the exact nature of this polycrystalline material.

Some bone cells, especially in the mandible, are known to mineralize during the life of an individual [6–8]. Their ultrastructure survives sufficiently well to enable organelles to be identified with TEM. Osteocytes which have mineralized in this way appear to have done so in an apoptotic state through a chemical process that preserves cell membranes, cytoskeleton and even nucleic bodies [8]. Furthermore, mineralized osteocytes have been identified in the fossil record in mammal bone aged up to 5 million years old [8]. Bacteria are also known to calcify. In dental calculus from humans as well as pigs, intracellular and intercellular mineralization has been observed [9, 10]. The remnant bacteria-like forms observed in this study appear to have some preserved cell ultrastructure, still detectable with TEM, and this supports their identification as mineralized bacteria. Whether these bacteria invaded the dentine tubule system at the time of death, when there is decomposition and proliferation especially of many gut bacteria that invade bones and teeth, or whether they have come from soil at some more recent stage [11–14], is uncertain. The bacterial forms illustrated in this study are not dissimilar to *Clostridia*, which is present in both the gut and soil [14]. To have penetrated dense fossil material or even highly mineralized tooth tissue from soil at some stage after death seems less likely than that they proliferated within the dentine tubule system in the immediate postmortem period. The fact that they are present at all in fossil material, opens a new window of possibilities for exploring the postmortem and diagenetic changes that occur in teeth and bones with TEM.

Another important finding of this study is that something resembling collagen may somehow be preserved. While it may not actually be collagen itself, some impression or replica of the collagen cross-banding pattern still appears to be present in demineralized bone matrix.

The fragment of enamel reveals several interesting features. The first is that there had clearly been some demineralization to the outer enamel, which shows as a subsurface zone of more porous enamel overlying a zone of more dense enamel. It is likely this occurred as the fossil was exposed on the surface a relatively short period of time before it was collected. Microstructural preservation in fossil enamel is often excellent and even better than in some modern teeth when seen in transmitted light microscopy. This is perhaps because some demineralization usually enhances incremental markings [15]. The cross-striation spacings measured in KNM-ER 6081 almost certainly diagnose it as belonging to *Paranthropus boisei*. Lacruz et al. [16] have documented the range of daily enamel formation rates in many early hominids and the only specimens where outer rates approach 7 µm per day belong to the genus *Paranthropus*.

Conclusions

Overall, we conclude that even after millions of years the microscopic preservation of fossil material can be extraordinary. Besides the preservation of normal hard tissue microstructure in KMN-ER 1817, there is considerable evidence for bacterial invasion of the dentine tubule system at some time, presumably after death. We also present preliminary evidence that at least remnants, or perhaps a replica of some kind, of collagen ultrastructure in bone can be preserved in fossil tissues as ancient as this. The low sums for the electron microprobe analysis indicate that there may be some organic matrix present in this fossil, and this view is supported by recent findings of Schweitzer et al. [17, 18] who demonstrated the presence of organic material in *Tyrannosaurus rex* bone from the Cretaceous. Thus, the minerals in hard tissues may protect parts of the organic matrix from degradation. This effect may even be enhanced as the fossilization process proceeds with even more mineral deposition and a change in mineral composition to more resistant F-apatite. Further studies are needed to determine the nature of this possible organic matrix. The enamel fragment we examined showed sufficient detail to enable adjacent cross-striation spacings to be measured and which diagnose it as belonging to the megadont hominin genus *Paranthropus*.

Acknowledgements

We thank the Government of Kenya and the trustees of the National Museums of Kenya for permission to study precious fossil material in their care. We are especially grateful to Meave Leakey, Emma Mbua and Alan Walker for enabling and supporting this study. We thank Lynne Bell, Charles Greenblatt and Helen Donoghue for their helpful observations and comments. We are grateful to the referees for their helpful comments. Aspects of this study were funded by a grant from The Leverhulme Trust to M.C.D.

References

1 Furseth Klinge R, Dean MC, Gunnæs A, Leakey MG: Microscopic structure and mineral distribution in tooth and periodontal tissues in a robust australopithecine fossil hominid from Koobi Fora, Kenya; in Żądzińska E (ed): Current Trends in Dental Morphology Research. 13th International Symposium on Dental Morphology. Łodz, University of Łodz Press, 2005, pp 233–242.

2 Wood BA: Koobi Fora Research Project. Hominid Cranial Remains. Oxford, Clarendon Press, 1991, vol 4, pp 1–466.

3 Dean C, Leakey MG, Reid D, Schrenk F, Schwartz GT, Stringer C, Walker A: Growth processes in teeth distinguish modern humans from *Homo erectus* and earlier hominins. Nature 2001;414:628–631.

4 Stormer JC, Pierson ML, Tacker RC: Variation of F and Cl X-ray intensity due to anisotropic diffusion in apatite during electron microprobe analysis. Am Mineral 1993;78:641–648.

5 Kohn MJ, Schoeninger MJ, Barker WW: Altered states: effects of diagenesis on fossil tooth chemistry. Geochim Cosmochim Acta 1999;63:2737–2747.

6 Kingsmill VJ, Boyde A: Mineralisation density of human mandibular bone: quantitative backscattered electron image analysis. J Anat 1998;192:245–256

7 Kingsmill VJ, Boyde A: Variation in the apparent density of human mandibular bone with age and dental status. J Anat 2002;192:233–244.

8 Bell LS, Kayser M, Jones C: The mineralized osteocyte: a living fossil. Am J Phys Anthropol 2008;137:449–456.
9 Furseth R: Further observations on the fine structure of orally exposed and carious human dental cementum. Archs Oral Biol 1971;16:71–85.
10 Furseth R: A study of experimentally exposed and fluoride treated dental cementum in pigs. Acta Odont Scand 1970;28:833–850.
11 Bell LS, Skinner MF, Jones SJ: The speed of *post mortem* change to the human skeleton and its taphonomic significance. Forensic Sci Int 1996;82:129–140.
12 Bell LS, Elkerton A: Unique marine taphonomy in human skeletal material recovered from the Medieval warship Mary Rose. Int J Osteoarchaeol 2007;18:523–535.
13 Turner-Walker G: The chemical and microbial degradation of bones and teeth; in Pinhasi R, Mays S (eds): Advances in Human Palaeopathology. New York, Wiley, 2008, pp 3–29.
14 Jans MME, Nielsen-Marsh CM, Smith CI, Collins MJ, Kars H: Characterisation of microbial attack on archaeological bone. J Archaeol Sci 2004;13:87–95.
15 Li C, Risnes S: SEM observations of Retzius lines and prism cross-striations in human dental enamel after different acid etching regimes. Arch Oral Biol 2004;49:45–52.
16 Lacruz RS, Dean MC, Ramirez Rozzi F, Bromage TG: Megadontia, striae periodicity and patterns of enamel secretion in Plio-Pleistocene fossil hominids. J Anat 2008;213:148–158.
17 Schweitzer MH, Wittmeyer JL, Horner JR, Toporski JK: Soft-tissue vessels and cellular preservation in *Tyrannosaurus rex*. Science 2005;307:1952–1955.
18 Schweitzer MH, Wittmeyer JL, Horner JR: Soft tissue and cellular preservation in vertebrate skeletal elements from the Cretaceous to the present. Proc Biol Sci 2007;274:183–197.

M. Christopher Dean, PhD
Cell and Developmental Biology, University College London
Anatomy Building, Gower Street
London, WC1E 6BT (UK)
Tel. +44 2076790462, E-Mail ucgacrd@ucl.ac.uk

Koppe T, Meyer G, Alt KW (eds): Comparative Dental Morphology.
Front Oral Biol. Basel, Karger, 2009, vol 13, pp 36–37

Dental Morphology: An Introduction

Mark F. Teaford

Center for Functional Anatomy & Evolution, Johns Hopkins University School of Medicine, Baltimore, Md., USA

Since their inception, the International Symposia on Dental Morphology have aimed to promote communication and collaboration among researchers who share the common bond of working on teeth and/or dental function. A quick look through this volume will show that the 14th Symposium, held in Greifswald, Germany in August of 2008, certainly continued that tradition, as the range of paper topics was even greater than in previous Symposia.

'Dental morphology' has traditionally been a cornerstone of these meetings and this was no exception as 20 papers were presented as posters or podium presentations in Greifswald. However, a close look at the papers accepted for publication from these sessions reveals that the focus of this work has expanded significantly in the last 30 years. In essence, 'dental morphology' has grown to include far more than just studies of tooth shape, as the questions asked now run the gamut from functional biology to evolutionary biology [1, 2].

Ungar's first paper is a case in point. Analyses of microscopic wear patterns on teeth were in their infancy 30 years ago [3, 4], and the method of data collection (dental microwear texture analysis) did not even exist [5]. Still, the technique raises exciting questions about ecological differences and niche partitioning in modern taxa and the mechanisms of evolution in prehistoric taxa.

Similarly, the second chapter by Lucas and coworkers shows we have come a long way from early functional studies of enamel thickness [6], where correlations between enamel thickness and diet differences gave us glimpses of functional significance but no real clue about how such things might work. Now, building on pioneering work on the physical properties of foods and biomaterials, Lucas' group shows that hard or soft items cause different types of fracture in enamel, and that these effects are visible at surprisingly different points on the tooth crown.

Another traditional application of 'dental morphology' has involved the use of tooth size and tooth shape to help in the identification of forensic cases [7]. Clearly, the ultimate usefulness of such an approach hinges on how common the traits are and how consistently they reflect differences between populations. The chapter by Edgar offers a cogent caution on the value of features such as Carabelli's trait and the canine mesial ridge in such enterprises.

Of course, another area in which tooth morphology has customarily played a huge role is paleontology, as teeth are the most common anatomical elements in the mammalian fossil record. The paper by Takai's group successfully blends conventional descriptions of tooth morphology with intriguing paleobiological inference, as new material from a Miocene owl monkey is used to yield new insights into the origin and evolution of nocturnality in this taxon.

Finally, given the complex blend of papers at these meetings, in areas such as growth and development and the clinical aspects of dental morphology, some papers really span multiple topic areas. The last paper in this section, by Kondo et al., is certainly a relevant example. On the one hand, it documents a variant in macaque mandibular morphology that is perhaps reminiscent of one in humans. It then uses CT analyses to shed light on possible genetic and biomechanical causes of such features [8–9].

References

1 Mayr E: The Growth of Biological Thought, Diversity, Evolution, and Inheritance. Cambridge, Harvard University Press, 1982.
2 Bock WJ: Explanation in Konstruktionsmorphologie and evolutionary morphology; in Schmidt-Kittler N, Vogel K (eds): Constructional Morphology and Evolution. Berlin, Springer, 1990, pp 9–29.
3 Dahlberg AA, Kinzey WG: Etude microscopique de l'abrasion et de l'attrition sur la surface des dents. Bull Group Int Rech Sci Stomatol Odontol 1962;5:242–251.
4 Walker A, Hoeck HN, Perez L: Microwear of mammalian teeth as an indicator of diet. Science 1978;201:908–910.
5 Ungar PS, Brown CA, Bergstrom TS, Walkers A: Quantification of dental microwear by tandem scanning confocal microscopy and scale-sensitive fractal analyses. Scanning 2003;25:185–193.
6 Martin LB: Significance of enamel thickness in hominid evolution. Nature 1985;314:260–263.
7 Scott GR, Turner CG II: The Anthropology of Modern Human Teeth: Dental Morphology and Its Variation in Recent Human Populations. Cambridge, Cambridge University Press, 2000.
8 Hylander WL: The functional significance of primate mandibular form. J Morphol 1979;160:223–240.
9 Moorrees CFA, Osborne RH, Wilde E: Torus mandibularis: its occurrence in Aleut children and its genetic determinants. Am J Phys Anthropol 1952;10:319–330.

Mark F. Teaford, PhD
Center for Functional Anatomy and Evolution
Johns Hopkins University School of Medicine
1830 E. Monument St., Room 303
Baltimore, MD 21205 (USA)
Tel. +1 410 955 7034, Fax +1 410 614 9030, E-Mail mteaford@jhmi.edu

Koppe T, Meyer G, Alt KW (eds): Comparative Dental Morphology.
Front Oral Biol. Basel, Karger, 2009, vol 13, pp 38–43

Tooth Form and Function: Insights into Adaptation through the Analysis of Dental Microwear

Peter S. Ungar

Department of Anthropology, University of Arkansas, Fayetteville, Ark., USA

Abstract

Mammalian molar form is clearly adapted to fracture foods with specific material properties. Studies of dental functional morphology can therefore offer important clues about the diets of fossil taxa. That said, analyses of tooth form provide insights into ability to fracture resistant foods rather than the food preferences of individuals. Recent work suggests that specialized occlusal morphology can relate to either preferred foods, or to occasionally eaten fallback items critical for survival. This paper reviews dental microwear texture analysis, a new approach that can be used to infer fracture properties of foods eaten in life. High-resolution 3D point clouds of microwear surfaces are collected and analyzed using scale-sensitive fractal analyses. Resulting data are free from operator measurement error, and allow the characterization and comparison of within-species variation in microwear texture attributes. Examples given here include four extant primate species (two folivores and two hard object fallback feeders), and two fossil hominin taxa. All groups show at least some individuals with simple microwear surfaces that suggest a lack of consumption of hard and brittle abrasive foods during the last few meals. On the other hand, some hard object fallback specimens have very complex surfaces consistent with consumption of hard, brittle foods. The latter pattern is also found in one hominin species. These results suggest that dental microwear texture analysis can help us determine whether craniodental specializations in fossil species are adaptations to preferred foods, or to less often but still critical fallback items.

Mammalian tooth form clearly relates to function. In primates for example, folivorous species tend to have high cusps connected by long mesiodistal crests capable of shearing through tough leaves, whereas frugivores, especially those that consume hard objects, tend to have blunter teeth well-suited to crushing [1–3]. Despite these obvious associations, however, relationships are not always quite so simple. It cannot always be assumed for example, that just because a tooth is *capable* of fracturing foods with given material properties, that such foods are commonly eaten. As Kinzey [4], queried more than three decades ago: 'Is it possible that features of the dentition are selected for on bases other than the 'primary specialization'?' Perhaps, as he theorized, 'when a food item is critical for survival, even though not part of the primary specialization, it will influence the selection of dental features' [4].

By analogy, while the driver of a sports car capable of 200 km/h would not likely floor the accelerator around the Marktplatz in Greifswald, it is nice to know the horsepower is available for the autobahn if needed! Likewise, although craniodental specializations may allow an animal to break down mechanically challenging items, an individual might not take advantage of this ability very often if those specializations do not preclude the consumption of preferred items. Such might be the case for animals that are forced to 'fallback' periodically on fracture-resistant foods at times of resource stress, when mechanically weaker, favored items are unavailable.

This leads to the question: 'How do we know whether specialized occlusal morphology is adapted to fracture resistant preferred foods or to fallback items that, while consumed less frequently, are still critical for survival? To answer this question for fossil forms, we need evidence of what individuals in the past actually ate on a day-to-day basis rather than what they were capable of eating. We need evidence of food fracture properties that is independent of interpretations based on dental morphology, such as dental microwear.

Dental Microwear and Fallback Food Exploitation

Dental microwear is the study of microscopic scratches and pits that form on a tooth's surface as the result of its use. Different foods leave different patterns of microwear. Among primates, for example, a diet dominated by hard, brittle items, such as palm nuts, should produce pits as these foods are crushed between opposing occlusal surfaces (think of a hammer and anvil). In contrast, a diet dominated by tough foods, such as some leaves, should result in scratches as abrasives are dragged across facets that slide past one another during shearing (think of a pair of scissors). Primates that eat foods with intermediate fracture properties, or have mixed diets, should have intermediate microwear patterns [5]. Thus, while occlusal morphology offers insights into the *potential* of a tooth to initiate and propagate cracks through a given item, dental microwear can provide information about the material properties of foods actually eaten by an individual during its lifetime.

Further, samples of microwear of many individuals may offer insights into breadth and frequencies of food types consumed by a fossil taxon. If most specimens of a primate species with long molar shearing crests have linear, parallel striations on their microwear facets, for example, tooth form probably reflects a 'primary specialization'. If only a few of these surfaces are dominated by scratches, molar morphology more likely relates to a fallback food adaptation.

These expectations follow what has been termed the 'Last Supper' phenomenon [6]. Mammalian dental microwear 'turns over' as surfaces wear down and old features are replaced by new ones. If diet changes, so too does microwear patterning. The rate of turnover varies with diet and abrasives in or on food items eaten, but individual features often appear and disappear in a matter of days [7] (fig. 1). With a sufficiently large and representative sample then, we should be able to get a sense of the range of variation, and dietary preferences of a group over time.

Some specific predictions are illustrated in figure 1. If most individuals in a sample consumed tough foods for their last few meals, their microwear surfaces are expected to be dominated by striations. If most ate hard, brittle foods, their surfaces should cluster in the heavily pitted range. If individuals varied in the fracture properties of the foods they ate over time, their microwear patterns should to be scattered. The prediction for hard object fallback exploitation is most individuals with striated or mixed microwear surfaces, but some toward the heavily pitted extreme as well.

Microwear Texture Analysis

Assessment of within species variation requires a method for characterizing microwear that minimizes 'noise' in the system. Surfaces can have hundreds of small, overlapping features prone to measurement error [8, 9]. Errors are exacerbated in SEM based studies by inconsistencies in geometric relations of the electron source, subject

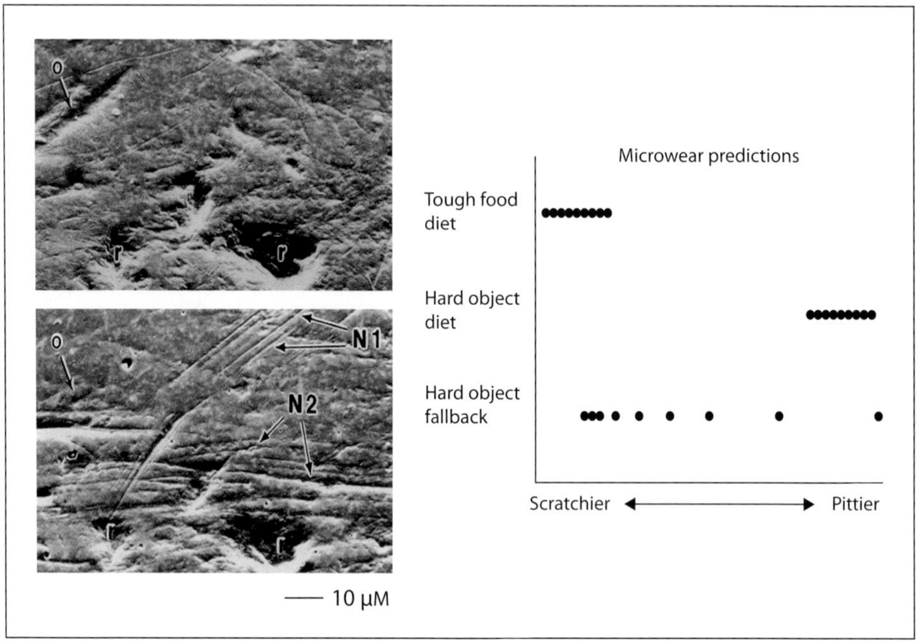

Fig. 1. Microwear turnover rates and within-species pattern predictions. Left = Microwear of a laboratory vervet monkey before and after 3 days (courtesy of Mark Teaford, see Teaford and Oyen [7], for details). Right = Predicted microwear patterning for samples with different diets.

and collector, which can change shading and apparent relief of a two-dimensional representation of a 3D microwear surface. In order for us to be confident that our data reflect the subtleties of within species variation then, we need a reduction in operator error that yields consistent repeatability of three-dimensional measurements. Dental microwear texture analysis was developed for this purpose [10–13]. Here I review some data from the literature [11–13] to illustrate the process.

Primate microwear texture analysis usually involves examination of 'phase II' facets of second molars, as these surfaces have been shown to effectively separate taxa by diet [14]. The procedure begins with an instrument capable of generating 3D point clouds with sufficient detail. Studies to date have used a Sensofar Plμ white-light scanning confocal profiler (Solarius Development Inc., Sunnyvale, Calif., USA) with a ×100 objective lens. Resulting data matrices each have a lateral sampling interval of 0.18 μm, a vertical resolution of 0.005 μm, and a field of view of 138 × 102 μm. Data are then collected for four adjoining fields to yield a total planimetric working area of 276 × 204 μm.

Individual scan data are then analyzed using Toothfrax (Surfract, www.surfract.com) and SFrax scale-sensitive fractal analysis (SSFA) software packages. SSFA operates on the principle that surface textures appearing smooth at a course scale can be rough at a finer scale (think of our racing car on the autobahn compared with an ant trying to cross the highway – to the car, the road will appear smooth, but to the ant it will appear exceptionally rough). Indeed, the apparent length of a profile across a surface, area, and volume of that surface all change with scale of observation.

Several aspects of surface texture can be characterized by considering change with scale. A

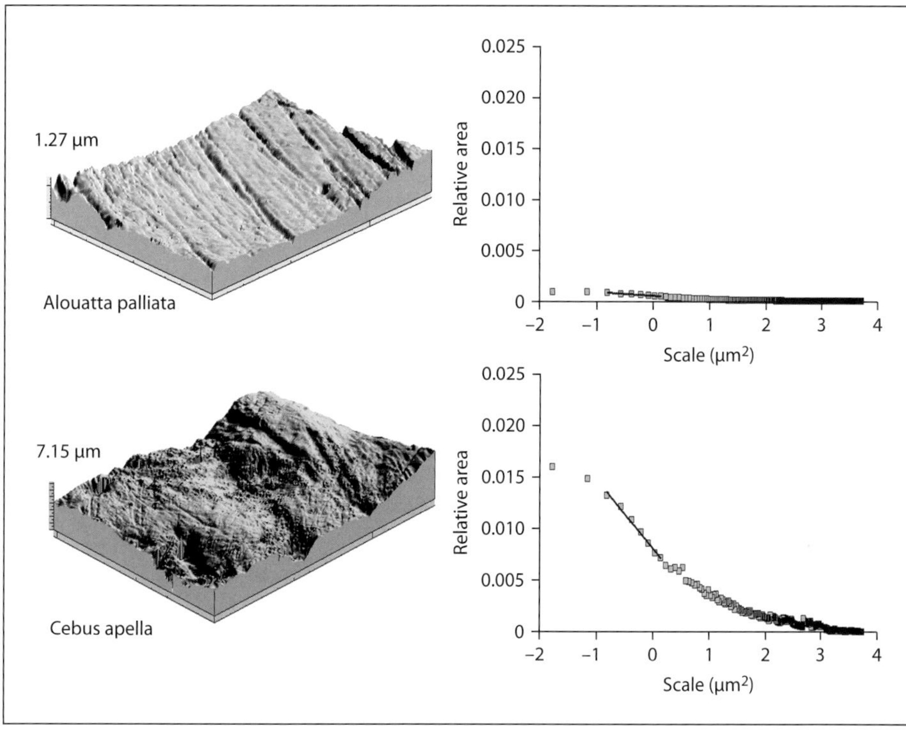

Fig. 2. Area-scale fractal complexity of dental microwear textures. Left = Three-dimensional axiomatic representations of point clouds for two specimens (102 × 139 µm each). Right = Corresponding *Asfc* plots for these two specimens.

surface that appears to increase progressively in relative area when measured at finer and finer scales, for example, is considered complex (fig. 2). Area-scale fractal complexity (*Asfc*) can be measured as the slope of the steepest part of the curve fit to a log-log plot of relative area over the scales at which measurements were made. High texture complexity is often recorded for heavily pitted surfaces, and has been suggested to be a good proxy for the consumption of hard, brittle foods [11–13]. Other texture attributes have also proven useful for distinguishing microwear surfaces, including anisotropy, heterogeneity, scale of maximal complexity, and textural fill volume. While space limitations prevent their consideration here, each is described in detail in the literature [12, 13].

Results and Discussion

Some complexity data compiled from published sources [11–13] are illustrated in figure 3. These data include high-resolution epoxy replicas [14, 15] of the second molars of wild-caught individuals representing two relatively folivorous extant species, *Alouatta palliata*, and *Trachypithecus cristatus*, and two taxa known to consume hard, brittle foods at times of resource scarcity [16], *Lophocebus albigena* and *Cebus apella*. Complexity data for two fossil hominin species, *Australopithecus africanus* and *Paranthropus robustus*, are also presented.

All groups have at least some specimens with low surface complexity values. This suggests a lack of consumption of hard and brittle abrasive

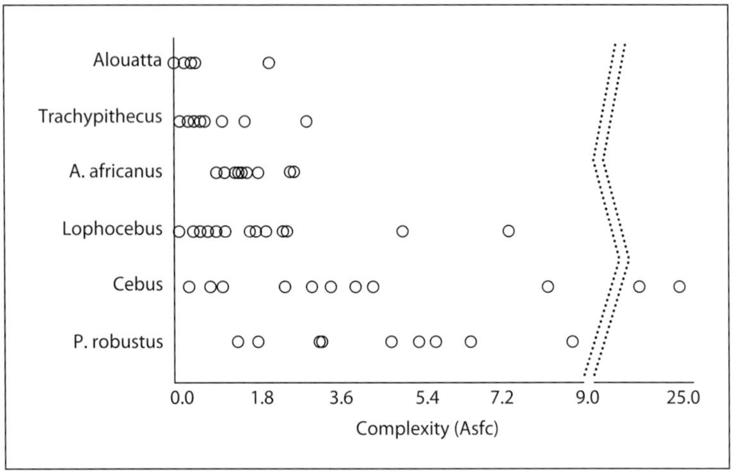

Fig. 3. Microwear texture complexity results for modern primates and fossil hominins.

foods by those individuals during their last few meals. In fact, all of the extant folivores fall into in this category. By contrast, the two groups known to consume hard, brittle foods have a few specimens with high levels of complexity. This is as expected for a hard object fallback feeder.

Results for the early hominins can be interpreted in this light. *Australopithecus africanus* has long been known to evince striated microwear surfaces whereas those of *Paranthropus robustus* are more pitted [6, 17]. This has been taken to indicate that *P. robustus* specialized on hard, brittle foods, as implied by its very robust craniodental toolkit [18]. Texture data presented here suggest that the story may actually be more complicated. First, there is overlap in microwear complexity between the samples. Low *Asfc* values suggest a lack of hard-brittle abrasive foods in the diets of the *A. africanus* and some *P. robustus* individuals. The pattern of a few high complexity values for *P. robustus*, on the other hand, is similar to patterns for *Lophocebus albigena* and *Cebus apella*. This is consistent with the hypothesis that the 'robust' australopiths occasionally consumed hard objects, not unlike grey-cheeked mangabeys or brown capuchins today.

If so, anatomical specializations of the *Paranthropus* craniodental toolkit may reflect a 'fallback' adaptation rather than selection for preferred food items. This interpretation mirrors recent work on stable isotopes and occlusal morphology of these hominins, as well as models derived from studies of living great apes [19–21]. It also follows a more general principle in ecology known as Liem's Paradox. As Robinson and Wilson [22] articulated, 'some resources are intrinsically easy to use and are widely preferred, while others require specialized phenotypic traits on the part of the consumer. This asymmetry allows optimally foraging consumers to evolve phenotypic specializations on nonpreferred resources without greatly compromising their ability to use preferred resources.'

Microwear can be especially useful for studies of fossils then, because dental functional morphology may not in and of itself tell us whether phenotypic specializations reflect preferred foods or non-preferred resources. It therefore holds the potential to inform us on the very nature of selection and adaptation for fracture resistant foods.

Acknowledgements

I thank Thomas Koppe, Kurt Alt, Georg Meyer, and Mark Teaford for their invitation to participate in and hospitality during the 14th International Symposium on Dental Morphology. I am also grateful to Robert Scott, Mark Teaford, Frederick Grine and Alejandro Pérez-Pérez for their help collecting data and developing the techniques described in this paper, and to the anonymous reviewers for their helpful comments. This work was funded by the US National Science Foundation.

References

1 Kay RF, Covert HH: Anatomy and behavior of extinct primates; in Chivers DJ, Wood BA, Bilsborough A (eds): Food Acquisition and Processing in Primates. New York, Plenum Press, 1984, pp 467–508.

2 Meldrum DJ, Kay RF: *Nuciruptor rubricae*, a new pitheciin seed predator from the Miocene of Colombia. Am J Phys Anthrop 1997;102:407–427.

3 Ungar PS, M'Kirera F: A solution to the worn tooth conundrum in primate functional anatomy. Proc Natl Acad Sci USA 2003;100:3874–3877.

4 Kinzey WG: Feeding behavior and molar features in two species of titi monkey; in Chivers DJ, Herbert J (eds): Recent Advances in Primatology. Behavior. New York, Academic Press, 1978, vol 1, pp 373–385.

5 Teaford MF: A review of dental microwear and diet in modern mammals. Scanning Microsc 1988;2:1149–1166.

6 Grine FE: Dental evidence for dietary differences in *Australopithecus* and *Paranthropus:* A quantitative analysis of permanent molar microwear. J Hum Evol 1986;15:783–822.

7 Teaford MF, Oyen OJ: In vivo and in vitro turnover in dental microwear. Am J Phys Anthrop 1988;75:279.

8 Rose JC, Ungar PS: Gross wear and dental microwear in historical perspective; in Alt KW, Rosing FW, Teschler-Nicola M (eds): Dental Anthropology: Fundamentals, Limits, Prospects. Stuttgart, Gustav-Fischer, 1998, pp 349–386.

9 Grine FE, Ungar PS, Teaford MF: Error rates in dental microwear quantification using scanning electron microscopy. Scanning 2002;24:144–153.

10 Ungar PS, Brown CA, Bergstrom TS, Walkers A: Quantification of dental microwear by tandem scanning confocal microscopy and scale-sensitive fractal analyses. Scanning 2003;25:185–193.

11 Scott RS, Ungar PS, Bergstrom TS, et al: Dental microwear texture analysis reflects diets of living primates and fossil hominins. Nature 2005;436:693–695.

12 Scott RS, Ungar PS, Bergstrom TS, et al: Dental microwear texture analysis: technical considerations. J Hum Evol 2006;51:339–349.

13 Ungar PS, Scott RS, Scott JR, Teaford MF: Dental microwear analysis: historical perspectives and new approaches; in Irish JD, Nelson GC (ed): Dental Anthropology. Cambridge, Cambridge University Press, 2007.

14 Beynon, AD: Replication technique for studying microstructure in fossil enamel. Scanning Microsc 1987;1:663–669.

15 Teaford MF, Oyen OJ: Live primates and dental replication: new problems and new techniques. Am J Phys Anthropol 1989;80:73–81.

16 Lambert JE: The biology and evolution of ape and monkey feeding; in Henke W, Tattersall I (eds): Handbook of Paleoanthropology: Principles, Methods, and Approaches. Berlin, Springer-Verlag, 2007, pp 1207–1234.

17 Grine FE: Trophic differences between 'gracile' and 'robust' australopithecines: a scanning electron microcope analysis of occlusal events. S Afr J Sci 1981;77:203–230.

18 Robinson JT: Prehominid dentition and hominid evolution. Evolution 1954;8:324–334.

19 Sponheimer M, Passey BH, de Ruiter DJ, Guatelli-Steinberg D, Cerling TE, Lee-Thorp JA: Isotopic evidence for dietary variability in the early hominin *Paranthropus robustus*. Science 2006;314:980–982.

20 Ungar PS: Dental topography and human evolution: with comments on the diets of *Australopithecus africanus* and *Paranthropus robustus*; in Bailey S, Hublin JJ (eds): Dental Perspectives on Human Evolution: State of the Art Research in Dental Anthropology. New York, Springer-Verlag, 2007, pp 321–344.

21 Wrangham R: The cooking enigma; in Ungar PS (ed): Evolution of the Human Diet: The Known, the Unknown and the Unknowable. New York, Oxford University Press, 2007, pp 308–323.

22 Robinson BW, Wilson DS: Optimal foraging, specialization, and a solution to Liem's paradox. Am Nat 1998;151:223–235.

Peter Ungar, Professor and Chair
Department of Anthropology, University of Arkansas
Old Main 330
Fayetteville, AR 72701 (USA)
Tel. +1 479 575 6361, Fax +1 479 575 2508, E-Mail pungar@uark.edu

Koppe T, Meyer G, Alt KW (eds): Comparative Dental Morphology.
Front Oral Biol. Basel, Karger, 2009, vol 13, pp 44–48

Primate Dental Enamel: What It Says about Diet

Peter W. Lucas[a] · Paul J. Constantino[a] · James J.-W. Lee[a] · Adam Hartstone-Rose[b] · Herzl Chai[e] · Wah-Keat Lee[c] · Nathaniel Dominy[d]

[a]Department of Anthropology, George Washington University, Washington, D.C., [b]Department of Biology, Pennsylvania State University, Altoona, Pa., [c]Advanced Photon Source, Argonne National Laboratory, Argonne, Ill., and [d]Department of Anthropology, University of California Santa Cruz, Santa Cruz, Calif., USA; [e]School of Mechanical Engineering, Faculty of Engineering, Tel Aviv University, Tel Aviv, Israel

Abstract

What kinds of fractures do teeth sustain and how do they resist disintegration? This study involved the mechanical loading of extracted human and sea otter teeth using hard and soft indenters to simulate hard and soft diets. The tests were accompanied by real-time imaging. At least three types of fracture were seen in the enamel – median, radial and margin cracks. Each kind of fracture appears to have a different cause, although the distinction between median and radial cracks blurs as they propagate. Only margin cracks appear to form under soft indenters. Several aspects of tooth form can be described as devices to limit damage to a tooth crown against the onslaught of hard or soft foods. The damage modes of teeth are paralleled by the behavior of some bilayered hard foods.

Writing in the shorthand necessitated by brief reviews such as this, feeding by primates is something of a fight. Their teeth contact food particles aiming to fracture them, but in so doing they must try themselves not to be damaged. Foods resist this fracture because the plant and animal tissues that end up inside primate mouths are, with the exception of fruit flesh, not intended to be eaten. However, mechanical tooth-food interactions are repeated several thousands of times daily over a lifetime, so dental injuries are inevitable.

In current research, we are asking what kinds of fracture teeth sustain. It is well known that teeth do in fact accumulate cracks [1], but we ask here how teeth resist cracks becoming so extensive that they lead to failure (i.e. loss of tooth function). In this brief review, we indicate some of the dangers to which teeth are exposed and highlight the importance of structures in tooth crowns that seem to help maintain tooth integrity. We studied human and sea otter teeth because of the convergent molar morphology of these mammals [2]. We also indicate parallels in the structure of some hard foods, and suggest that some of the mechanisms they employ to resist disintegration are shared with teeth. The methodology required for this analysis is that of contact and fracture mechanics, familiar to materials scientists [3].

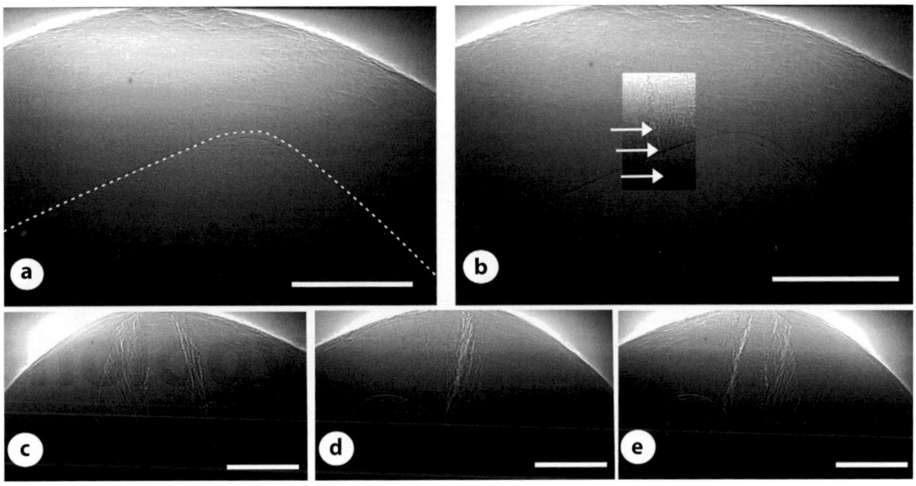

Fig. 1. Synchrotron images of damage to a lower molar cusp of a sea otter *(Enhydra lutris)*, caused by loading with a flat tungsten carbide disk, which has permanently flattened the cusp tip. **a** Image of the cusp during loading, but prior to fracture. The dotted white line indicates the position of the EDJ. **b** Initiation of a crack (arrowed) close to the EDJ. By rotating the cusp about its center of axis, images C-E show that the fully developed radial crack that develops from that in **b** is restricted to the enamel. Scale bar = 0.5 mm.

Materials and Methods

Upper and lower extracted human teeth were obtained by J.L. from the laboratory of the American Dental Association located at the National Institute of Standards and Technology (NIST), Gaithersburg, and testing approval was granted by the NIST IRB board. These teeth came from 18- to 25-year-old male and female patients. Teeth from deceased sea otters came from Jim Estes (University of California Santa Cruz) and Nate Dominy, with permission from Melissa Miller (California Department of Fish and Game). All teeth were kept wet and examined for damage prior to testing. The crowns of extracted teeth (human premolars and third molars; sea otter first molars) were then loaded on the cusp tips in a mechanical tester by flat metal (steel or tungsten carbide) or polytetrafluoroethylene (PTFE or 'Teflon') punches. Damage to specimens was observed in real time either by a high-resolution camera linked to a video recorder or in the Argonne Advanced Photon Source (APS) synchrotron. With the latter, the initiation of cracks with respect to the enamel-dentine junction (EDJ) could be determined (fig. 1a). After testing, teeth were sectioned and polished down to a 1 μm finish to understand better how cracking observed in the real-time images had progressed within the crown.

Results

A consistent finding when teeth were fractured during tests was that cracking took place within the enamel. This was evident both as phase contrast synchrotron images (fig. 1) and in light micrographs of sectioned and polished teeth. Damage to dentine was always a secondary event. Under soft indenters, the load required to initiate cracking in the enamel was much higher than that under a hard indenter.

Three different modes of fracture were observed (fig. 2). In human teeth, 'median' cracks initiated from the plastic zone that forms immediately underneath the indenter [3, 4], and then spread down the crown (fig. 2a). In contrast, in sea otter teeth, 'radial' cracks initiated at the EDJ, from flexure of the stiff enamel on a soft dentin support [5–7]. These two forms of cracking were only seen with a hard indenter. With both hard and soft indenter types, another fracture location

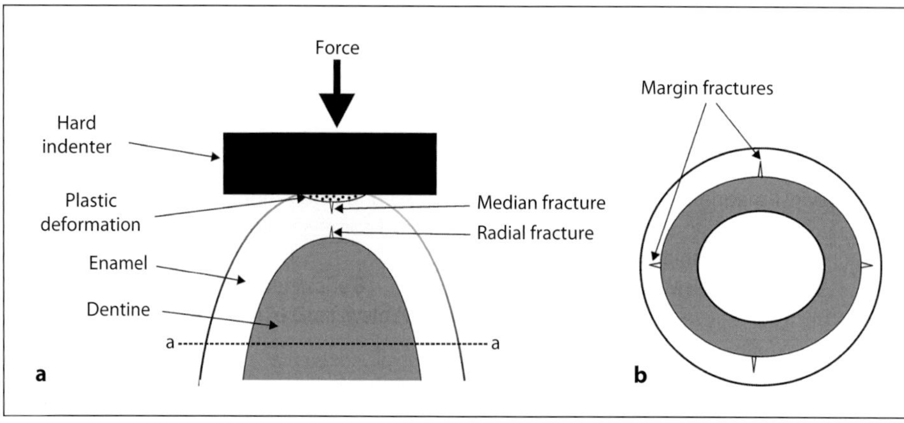

Fig. 2. Three types of cracking shown schematically for a bunodont tooth. **a** Two possibilities. Median faults develop below a plastic zone induced by a very hard indenter, while radial fractures originate at the EDJ. Though the origin of both fractures is different, they probably propagate through the tooth in similar fashion. **b** The line a-a shows the position of the transverse section of the crown. Here, deep flaws called margin cracks are shown initiating again at, or near, the EDJ.

was seen in both humans and sea otters. Margin cracks originated close to the cement-enamel junction, and then spread up the crown towards the cusp tip (fig. 2b). With soft indenters, margin fractures predominated [8].

Discussion

Despite their different origins, a general equation governs the conditions for both median and radial cracks once they have developed sufficiently to cause catastrophic failure (fig. 3) [9]. When failure for a tooth cusp is defined as 'fracture that extends down to the cervical margin', the failure force is:

$$P_F = B_F T d^{1.5} / \log_{10}(E_{enamel}/E_{dentine}) \qquad (1)$$

where T is the enamel toughness, d is the enamel thickness, E_{enamel} is the enamel modulus, and $E_{dentine}$ is the modulus of the encapsulated dentine [6]. The coefficient B_F is $[13.5 + 2.1(r_e/d)]$, where r_e is the effective radius of contact between contacting surfaces [9]. This equation leads immediately to design considerations for teeth against a hard diet. Against hard food particles, tooth cusps should have a high radius of curvature, i.e. be blunt, because this increases B_F, which will raise the force at which cracks become catastrophic. However, greater protection can be obtained from thickening the enamel, i.e. by increasing d, because the break force is proportional to $d^{1.5}$. Thus, teeth designed to process a hard diet should be blunt-cusped and thick enameled. A significant increase in the force to propagate cracks can also be obtained by manipulating enamel microstructure, it having recently been shown that decussation of the enamel leads to higher toughness by frustrating crack opening [10]. The positioning of Hunter-Schreger bands in the enamel may depend on whether the origin of the cracks is relatively superficial (median) or deep (radial).

As an example of the use of equation 1, we apply it to the cusp of the lower molar of a sea otter with a radial crack (fig. 1). For this cusp, the load

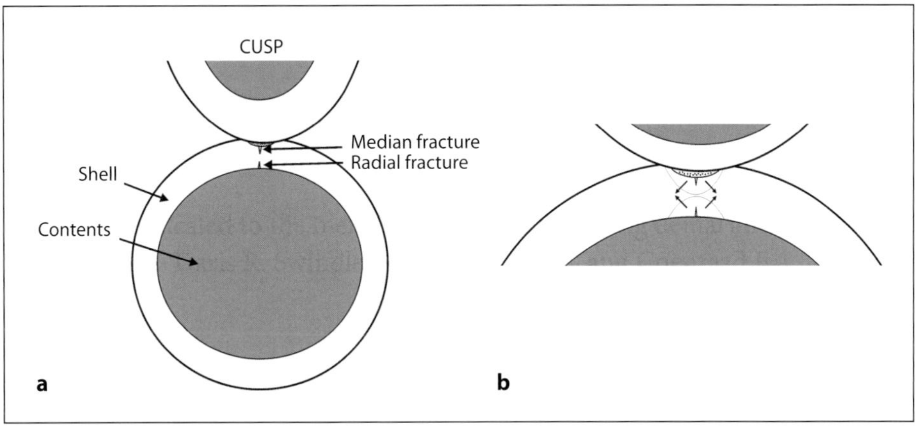

Fig. 3. a Contact of a bunodont cusp with a bilayered 'hard' food (much less hard than enamel). This can cause median or radial cracks depending on the contact. **b** Once either a median or radial crack is initiated, the growth becomes increasingly similar as they spread around the food shell.

was 311 N, while 'plumb line' enamel thickness d, measured in the direction of the load, was 0.5 mm. Assuming 'human' values for the moduli for sea otter enamel (E_{enamel} = 90 GPa) and dentine ($E_{dentine}$ = 20GPa), the logarithm of the ratio of these properties being much less sensitive to error than toughness, then taking $B_F \approx 19$ (i.e. with r_e = 1.5 mm), T is predicted as 0.97 MPa m$^{1/2}$, which is similar to the value for human enamel, and entirely reasonable as a 'ballpark' estimate.

Median cracks can only initiate after enamel has plastically deformed. This type of deformation is only possible when enamel is contacted by particles in the diet that are at least 40% as hard and stiff as that dental tissue. One of the major reasons for the high stiffness and hardness of dental enamel is surely to avoid this possibility because plastic yielding can take place at very low forces [7]. Particles of extraneous grit (quartz) that enter the mouth are hard enough to make enamel yield and the subsequent microwear is the product of plastic deformation. However, grit particles, as also plant phytoliths, are far too small to initiate median fractures in the enamel because a size effect, the 'brittle-ductile' transition, comes into play [7]. The larger particles of the actual foods that primates eat are very rarely harder or stiffer than enamel [11], so median cracks in the enamel are going to be rare. However, when their teeth bite on bilayered 'hard' foods such as seeds [12, 13], in which an inedible shell protects a nutritious core, both median and radial cracks in the seed shell can often be seen (fig. 3a).

With a soft diet, margin cracks in teeth are much more important than cracking at the region of the cusp under load. This is because of compression that suppresses local cracking. However, cracks can still be generated much lower down the tooth crown towards the cement-enamel junction due to 'hoop' stresses produced as the tooth bulges [8]. Margin fractures in teeth appear to fail at a force proportional to the product $r_t d^{0.5}$, where r_t is the radius of the tooth [8]. Here, thickening the enamel will have much less effect in resisting cracks than for median-radial cracks. Instead, the equivalent of a barrel hoop placed round the crown, i.e. a cingulum, may be more effective because it effectively increases r_t [7].

Acknowledgements

Many thanks go to Mark Wagner and Will Morton (George Washington University) for their technical help.

Use of the Advanced Photon Source at Argonne National Laboratory was supported by the US Department of Energy, Office of Science, Office of Basic Energy Sciences, under Contract No. DE-AC02-06CH11357.

References

1 Sognnaes RF: The organic elements of enamel. IV. The gross morphology and the histological relationship of the lamellae to the organic framework of the enamel. J Dent Res 1950;29:260–269.

2 Walker AC: Diet and teeth: dietary hypotheses and human evolution. Phil Trans R Soc Lond [B] 1981;292:57–76.

3 Lawn BR: Fracture of Brittle Solids, ed 2. Cambridge, Cambridge University Press, 1993.

4 Lawn BR, Evans AG: A model for crack initiation in elastic/plastic indentation fields. J Mater Sci 1977;12:2195–2199.

5 Chai H, Lawn BR, Wuttiphan S: Fracture in brittle coatings with large interlayer modulus mismatch. J Mater Res 1999;14:3805–3817.

6 Rhee Y-W, Kim H-W, Deng Y, Lawn BR: Contact-induced damage in ceramic coatings on compliant substrates. J Am Ceram Soc 2001;84:1066–1072.

7 Lucas PW, Constantino P, Wood BA, Lawn BR: Dental enamel as a dietary indicator in mammals. Bioassays 2008;30:374–285.

8 Chai H, Lee J J-W, Kwon J-Y, Lucas PW, Lawn BR: A simple model for enamel fracture from margin cracks. Acta Biomat 2009;5:1663–1667.

9 Lawn BR, Lee J J-W: Analysis of fracture and deformation modes in teeth subjected to occlusal loading. Acta Biomat 2009;5:2213–2221.

10 Bajaj D, Nazari A, Eidelman N, Arola DD: A comparison of fatigue crack growth in human enamel and hydroxyapatite. Biomaterials 2008;29:4847–4854.

11 Lucas PW, Constantino PJ, Chalk J, Ziscovici C, Wright BW, Fragaszy DM, Hill DA, Lee J J-W, Chai H, Darvell BW, Lee PKD, Yuen TDB: Indentation as a technique to assess the mechanical properties of fallback foods. Am J Phys Anthrop. In press.

12 Lucas PW, Lowrey TK, Pereira B, Sarafis V, Kuhn W: The ecology of *Mezzettia leptopoda* Hk. f. et Thoms. (Annonaceae) seeds as viewed from a mechanical perspective. Func Ecol 1991;5:345–353.

13 Williamson L, Lucas PW: The effect of moisture content on the mechanical properties of a seed shell. J Mater Sci 1995;30:162–166.

Dr. Peter W. Lucas
Department of Anthropology, George Washington University
2110 G St NW
Washington DC 20052 WA (USA)
Tel. +1 202 994 6964, Fax +1 202 994 6097, E-Mail peterwlucas@gmail.com

Koppe T, Meyer G, Alt KW (eds): Comparative Dental Morphology.
Front Oral Biol. Basel, Karger, 2009, vol 13, pp 49–54

Testing the Utility of Dental Morphological Traits Commonly Used in the Forensic Identification of Ancestry

Heather J.H. Edgar

University of New Mexico, Albuquerque, N. Mex., USA

Abstract

Few human variants are truly population specific, with 100% frequency in one group and 0% in others. However, for traits to be of use in forensic identification they must be as specific to a population as possible. Forensically, several dental morphological traits have been described as useful for determining an unknown individual's ancestry. For these traits to be of value, they should occur in their associated group in proportions statistically different from all other groups. Furthermore, ancestral groups not associated with the trait should have no significant frequency differences among them. To test this, frequencies of dental morphological traits listed in the forensic literature as useful for ancestry determination were compared among samples of African Americans, European Americans, Hispanic Americans, and Native Americans (n = 1625). χ^2 tests were conducted on dichotomized frequencies of ten trait observations, including incisor shoveling, Carabelli's trait, canine mesial ridge, and cusp seven. Results were mixed. For example, Native Americans have statistically different frequencies of shovel shaping from all other groups. However, statistical differences are seen among other groups as well. Only canine mesial ridge was consistently different in African Americans and not different among the other three groups. Unfortunately, this trait is not common in any group (2–21%), so lack of the trait is not indicative of ancestry. Overall, these commonly used traits may not be of much actual value in ancestry determination. Combining these traits with other observations and more sophisticated statistical tools may be of more practical value.

Forensic anthropology is a specialty within forensic medicine concerned with the assessment of human skeletal remains and their environments within a legal context [1]. The scope of the field includes analysis of dental remains for estimation of age, sex, ancestry, and individual characteristics. Dental morphology, in the anthropological sense, involves observations of the minor variations in the cusps, ridges, grooves, and root structures that in general make up the surfaces of teeth [2]. Dental anthropologists generally utilize many dental characteristics and relatively complex statistics to describe populations. When using dental morphology, however, forensic anthropologists tend to use a very limited number of traits, with simple statistical or qualitative approaches to assign an individual to a population.

It has long been known that dental morphological variation could have forensic applications, especially in the estimation of ancestry [3]. A few dental traits have come to be part of the arsenal of morphological characteristics on which forensic anthropologists rely. These especially include two very well known traits; shovel-shaped incisors and Carabelli's trait. However, the actual utility of these traits has not been formally tested. Additionally, while nearly all practitioners of forensic anthropology are familiar with this limited subset of dental variants, few have ventured beyond them to explore the value of lesser-known traits. In fact, while a recent review of forensic applications of dental analyses listed many papers that indicated the value of morphological characteristics in interpreting population variation, the author only cited two papers that specifically addressed the value of morphological analysis in determining the ancestry of an unknown individual [4–6].

The research presented here is part of an ongoing effort to increase and improve the use of dental morphology in forensic anthropology. I provide a test of the efficacy of the standard application of dental variation in medicolegal contexts by asking: Do traditional dental morphology 'markers' of ancestry really work?

What Traits Are Forensic Anthropologists Using?

Lasker and Lee [7] authored one of the first surveys in English of the use of dental characteristics to determine ancestry in a forensic setting. They noted that shovel-shaped incisors are most common in persons of Asian descent, and that Carabelli's trait is most common in persons of European descent. They did not identify any traits that were more common in persons of African ancestry. Shovel-shaped incisors and Carabelli's trait remain the most common, if not the only, dental traits used in forensic analyses [1, 8–10]. However, recent research [11, 12] has shown that Carabelli's trait frequencies are variable in all worldwide populations, and not particularly useful in many populations.

In addition to these two traits commonly described in the literature for estimating ancestry in an individual, I found references to a few other traits, especially cusp 7 (a cusp occurring in the lingual groove between the metaconid and entoconid), which is considered to indicate recent African ancestry [6, 13] as is canine mesial ridge, or Bushman canine [14]. A few other traits were occasionally described as useful, including multi-cusped premolars and molar complexity. However, attribution was inconsistent [3, 15] and scoring methods were not provided. The current report presents results concerning the two most common traits mentioned in the literature, shovel-shaped incisors and Carabelli's trait, as well as two less commonly noted traits, cusp seven and canine mesial ridge. Currently, no dental morphological traits have been associated with Hispanic Americans, who result from the admixture of European and African derived populations with Native Americans. It is expected that their dental morphological trait frequencies should be intermediate among those of the populations from which they descend.

Materials and Methods

Materials
Due to secular changes in populations, methods developed for medicolegal applications should be tested on living or very recent samples. Therefore, all of the materials used in this study date to the second half of the 20th century. Samples are listed in table 1, and include materials representing contemporary African, European, Hispanic, and Native Americans.

Methods
Observations were made of traditional forensic dental morphology 'markers' of ancestry. Shovel shaping, which is associated with Asian ancestry and likely to be associated with Native Americans, was observed on all eight adult incisors and the two maxillary canines. Canine mesial ridge, associated with African ancestry,

Table 1. Samples used in this study

Institute	Collection	African American	European American	Hispanic American	Native American	n
Arizona State University (Nichol, 1990)	Dahlberg collection				434	434
Arizona State University (Nichol, 1990)	Saint John's Indian School, Ariz., USA				194	194
Arizona State University (Nichol, 1990)	Keams Canyon Public Health Service Hospital, Ariz., USA				533	533
Arizona State University (Edgar, 2005)	University of Washington	10				10
Case Western Reserve University (Edgar, 2005)	Bolton-Brush		54			54
University of Tennessee Memphis Health Center (Edgar, 2005)	dental casts	100	101			201
University of New Mexico	documented skeletons			20		20
University of New Mexico	Hixon dental casts			38		38
University of New Mexico	Economides dental casts	5	50	86		141
Total		115	205	144	1,161	1,625

was observed on the maxillary permanent canines. Carabelli's trait, which is associated with European ancestry, was observed on the first and second maxillary permanent molars. Cusp seven was observed on the first and second mandibular adult molars. Left and right observations were combined using the expression count method [16]. Scores were dichotomized following Scott and Turner [2].

In this test, a trait is considered useful in the forensic estimation of ancestry if it meets two conditions:

(1) The frequency in the group associated with the trait should be statistically significantly different than in other groups.

(2) There should be no statistically significant differences in the frequencies of the trait among the unassociated groups.

Statistical significance was tested using χ^2 tests.

Results

All results are shown in table 2. For 'Asian' traits, shovel-shaped incisors, Native Americans are consistently different from African and European Americans. However, there are also significant differences between African Americans and European Americans on every tooth on which the trait was observed. For Carabelli's trait, associated with European ancestry, there is a great deal of variation among the three ancestral groups. For the first molar, all groups show significant differences except for the African American/European American comparison. For

Table 2. Trait expression frequency differences (below the diagonal) and associated χ^2 significances (above the diagonal)

'Asian' traits		AA	EA	HA	NA
UI1 shovel-shaped	AA		**9.79**	**7.3**	**11.12**
	EA	0.18		*0.01*	**62.28**
	HA	0.18	0		**37.88**
	NA	0.19	0.37	0.37	
UI2 shovel-shaped	AA		*0.73*	*1.27*	**167.78**
	EA	0.02		**4.23**	**233.6**
	HA	0.03	0.05		**119.2**
	NA	0.65	0.67	0.62	
UC shovel-shaped	AA		**8.98**	*0.41*	**24.64**
	EA	0.08		**12.63**	**61.89**
	HA	0.03	0.11		**13.05**
	NA	0.23	0.31	0.2	
LI1 shovel-shaped	AA		**19.54**	**4.91**	**578.03**
	EA	0.17		**5.05**	**492.03**
	HA	0.06	0.11		**409.8**
	NA	0.91	0.74	0.85	
LI2 shovel-shaped	AA		*0.21*	**16.75**	**236.98**
	EA	0.02		**43.5**	**591.14**
	HA	0.42	0.4		**133.9**
	NA	0.82	0.8	0.4	

'African' traits		AA	EA	HA	NA
UC mesial ridge	AA		**23.81**	**9.71**	**83.13**
	EA	0.19		*0.89*	*0.01*
	HA	0.17	0.02		*1.81*
	NA	0.19	0	0.02	
LM1 cusp 7	AA		**30.71**	*0.34*	**146.27**
	EA	0.07		**18.86**	**7.99**
	HA	0.04	0.27		**83.44**
	NA	0.38	0.07	0.34	
LM2 cusp 7	AA		**8.3**	**17.89**	**42.46**
	EA	0.07		**48.76**	*0.34*
	HA	0.27	0.34		**242.68**
	NA	0.7	0.01	0.35	
'European' traits					
UM1 Carabelli's	AA		*0.01*	**4.59**	**29.66**
	EA	0.01		**5.88**	**42.19**
	HA	0.07	0.15		**5.19**
	NA	0.27	0.28	0.13	
UM2 Carabelli's	AA		*3.54*	*0.92*	**103.39**
	EA	0.1		*0.37*	**55.79**
	HA	0.07	0.03		**49.85**
	NA	0.25	0.15	0.18	

Significant χ^2 results are in bold; nonsignificant results are in italics.

the second molar, only Native Americans are significantly different than the other groups. This lack of patterning is consistent with prior work on the characteristic by Hawkey and Turner [11]. There are two traits tested for association with African ancestry. The expression of canine mesial ridge meets both of the conditions for utility described above. Cusp seven meets neither, as African Americans are not significantly different from Hispanic Americans for the first molar, and there are significant differences between Hispanic Americans and European and Native Americans for the second molar. Unfortunately, canine mesial ridge is only seen in 2–21% of all worldwide populations [2], so the lack of the trait in any particular individual's dentition is not necessarily indicative of ancestry.

Discussion

This first formal test of the dental characteristics commonly used in forensic anthropology shows that, in fact, the traits are of limited utility. This finding indicates that forensic researchers should utilize an approach common among dental anthropologists, drawing on many characteristics rather than a few. A recent report [17] makes this point. In estimating ancestry, the authors considered a number of dental morphological traits including shovel shape, but also enamel extensions and molar and premolar cusp numbers. They estimated that the unknown individual was of Asian ancestry, specifically South Asian, due to a generally Sundadont dental pattern. Once identified, the individual was found to be from Vietnam.

What was missing from this report [17] was a statistical approach to the question of ancestry. The history of ancestry estimation in forensic anthropology has been mainly qualitative [1, 3, 7, 8, 15]. In the recent past, approaches using skeletal variation have become much more quantitative [18]. New approaches using dental morphology could use multiple traits as well as multiple degrees of expression (rather than dichotomization), coupled with more sophisticated statistics. A few papers have started this trend [5, 6], and have received attention [19, 20], indicating the potential value to the forensic science community.

Acknowledgements

I would like to thank the organizers of the Greifswald meeting of the International Symposium on Dental Morphology for their hard work and hospitality. My thanks also goes to the following organizations and individuals for their generosity with their dental cast collections: the University of Tennessee Health Sciences Center, Case-Western Reserve University, Arizona State University, James Economides, DDS and Linda Hixon, DDS.

References

1 Krogman WM, Iscan MY: The Human Skeleton in Forensic Medicine, ed 2. Springfield, Thomas, 1986.
2 Scott GR, Turner CG II: The Anthropology of Modern Human Teeth: Dental Morphology and Its Variation in Recent Human Populations. Cambridge, Cambridge University Press, 2000.
3 Kimminau KS: Dental variation and racial estimation: problems and practical forensic applications. Napa Bulletin 13: Race, Ethnicity and Forensic Anthropology, 1993, pp 68–73.
4 Schmit CW: Forensic dental anthropology: issues and guidelines; in Irish JD, Nelson GC (eds): Technique and Application in Dental Anthropology. Cambridge, Cambridge University Press, 2008, pp 266–292.
5 Edgar HJH: Prediction of race using characteristics of dental morphology. J Forensic Sci 2005;50:269–273.
6 Lease LR, Sciulli PW: Brief communication: discrimination between European-American and African-American children based on deciduous dental metrics and morphology. Am J Phys Anthropol 2005;126:56–60.
7 Lasker GW, Lee MMC: Racial traits in human teeth. J Forensic Sci 1957;2:401–419.
8 Burns KR: Forensic Anthropology Training Manual. Upper Saddle River, Prentice-Hall, 1999.
9 Pretty IA, Sweet D: A look at forensic dentistry. 1. The role of teeth in the determination of human identity. Br Dent J 2001;190:359–366.
10 Hinkes MJ: Shovel-shaped incisors in human identification; in Gill GW, Rhine S (eds): Skeletal Attribution of Race: Methods for Forensic Anthropology. Albuquerque, Maxwell Museum of Anthropology, 1990, pp 21–26.
11 Hawkey DE, Turner CG II: Carabelli's trait and forensic anthropology: whose teeth are these?; in Lukacs JR (ed): Human Dental Development, Morphology, and Pathology: A Tribute to Albert A. Dahlberg. Eugene, Department of Anthropology, University of Oregon, 1998, pp 41–50.
12 Correia A, Pina C: Tubercle of Carabelli: a review. Dental Anthropol J 2000;15:18–21.

13 Teisler V: New case of an African tooth decoration from colonial Campeche, Mexico. Homo 2002;52:277–282.
14 Irish JD, Morris DH: Technical note: canine mesial ridge (Bushman canine) dental trait definition. Am J Phys Anthropol 1996;99:357–59.
15 Rhine S: Non-metric skull racing; in Gill GW, Rhine S (eds): Skeletal Attribution of Race: Methods for Forensic Anthropology. Albuquerque, Maxwell Museum of Anthropology, 1990, pp 9–20.
16 Turner CG II: Expression count: a method for calculating morphological dental trait frequencies by using adjustable weighting coefficients with standard ranked scales. Am J Phys Anthropol 1985;68:263–267.
17 Kemkes-Grottenthaler A: The reliability of forensic osteology: a case in point: case study. Forensic Sci Int 2001;117:65–72.
18 Ousley SD, Jantz RL: FORDISC 3.0:Personal Computer Forensic Discriminant Functions. Knoxville, University of Tennessee, 2005.
19 Cattaneo C: Forensic anthropology: developments of a classical discipline in the new millennium. Forensic Sci Int 2007;165:185–193.
20 White TD, Folkens PA: The Human Bone Manual. New York, Academic Press, 2005.

Heather J.H. Edgar, PhD
Maxwell Museum of Anthropology
MSC01 1050, 1 University of New Mexico
Albuquerque, NM 87131 (USA)
Tel. +1 505 277 4415, Fax +1 505 277 1547, E-Mail hjhedgar@unm.edu

Koppe T, Meyer G, Alt KW (eds): Comparative Dental Morphology.
Front Oral Biol. Basel, Karger, 2009, vol 13, pp 55–59

Meaning of the Canine Sexual Dimorphism in Fossil Owl Monkey, *Aotus dindensis* from the Middle Miocene of La Venta, Colombia

Masanaru Takai[a] · Takeshi Nishimura[a] · Nobuo Shigehara[a] · Takeshi Setoguchi[b]

[a]Primate Research Institute, Kyoto University, Inuyama, and [b]Department of Geology and Mineralogy, Graduate School of Science, Kyoto University, Kyoto, Japan

Abstract

The owl monkey, *Aotus*, is the only modern nocturnal anthropoid with monogamous social structure. It has been demonstrated by the fossil species, *Aotus dindensis*, discovered from La Venta, Colombia, that the *Aotus* lineage had emerged as early as the middle Miocene (12–15 Ma). The type specimen of *A. dindensis*, which was discovered in 1986, preserves extremely large orbits, indicating a nocturnal habit. However, a few anatomical traits in living *Aotus*, such as the lack of a tapetum lucidum, indicates that nocturnality is a secondary adaptation from diurnal ancestry in this genus. Here we report new fossil specimens of *A. dindensis* from La Venta. The specimens include maxillary teeth and a mandibular fragment preserving lower molars. The detailed analysis of the specimen suggests that *A. dindensis* exhibits strong sexual dimorphism in the maxillary canine and premolars, which is traditionally associated with intense intermale competition for mates and/or food resources in non-monogamous, diurnal societies. As a result, the new fossil materials of *A. dindensis* demonstrate the first osteological evidence for the diurnal ancestry of the night monkey, *Aotus*. Moreover, the coexistence of large orbits and canine dimorphism suggests the presence of mosaic evolution in the craniodental characters of the *Aotus* lineage.

The evolution of nocturnality in the owl monkey is one of the most intriguing topics in modern primatology. The only known fossil record of the owl monkey is *Aotus dindensis*, which was discovered in the middle Miocene sediments of La Venta, Colombia, in 1986 [1]. The type specimen of *A. dindensis* (IGM-KU 8601, a mandible preserving left I_1-M_3 and right I_1 and a left maxillary fragment preserving alveoli of M^{1-2} and lingual fragment of M^3), were nearly identical to extant *Aotus* both in size and morphology, and the structure of the maxillary bones indicated an extremely large orbit, suggesting a nocturnal activity pattern for *A. dindensis* [1]. The combination of these data demonstrated that the *Aotus* lineage acquired nocturnal habits as early as the middle Miocene, and had hardly changed its basic morphology in almost 15 million years [1]. By contrast, anatomical evidence from modern *Aotus*, such as the lack of a tapetum lucidum and the presence of a central retinal fovea [2–5] has been used to suggest that nocturnality is a secondary adaptation from diurnal ancestry in this lineage. Thus, the retention of nocturnal traits in *Aotus dindensis* suggests that any diurnal ancestry in the *Aotus* lineage extended back into at least the early Miocene. The new fossil specimen of *A. dindensis* reported here sheds additional light on this intriguing issue.

The new specimen was collected in 1998 in the La Venta badlands (03°13′70N, 5°10'83W), about 7 km south of the type locality of *A. dindensis*. Stratigraphically, the site corresponds to the lowest part of the Tatacoa Red Member of the Villavieja Formation, which is about 100 m higher than the type horizon [6, 7]. The fission track ages obtained from the Tatacoa Red Member are 13.6 to 12.6 Ma [8].[1] The new specimen (IGM-KU 98001, 98002) consists of fragments and isolated teeth of the upper and lower jaws, representing a single individual as all teeth are nearly the same size and there is no overlap in tooth classes. P_4-M_3 are nearly identical to those of the type specimen (IGM-KU 8601) and extant *Aotus* (fig. 1), and M^{2-3} and I_2 are also identical to those of extant *Aotus* except for their larger size, confirming the evolutionary conservation of incisor and molar morphology.

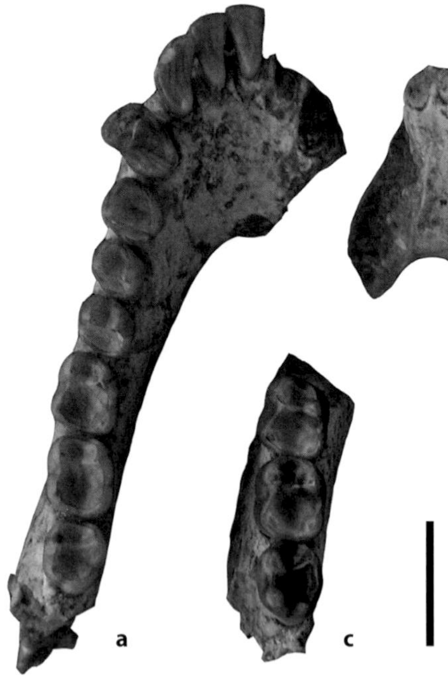

Fig. 1. Lower dentition of *A. dindensis*. **a** Type specimen (IGM-KU 8601, left mandibular corpus preserving left I_1-M_3 and right I_1). **b** New specimen, symphysial fragment (IGM-KU 98002). **c** left mandibular fragment preserving M_{1-3} (IGM-KU 98001). Note the morphological similarity especially in M_{1-3} between the two specimens. Scale bar = 5 mm.

In contrast to the strong similarities in the incisors and molars, the upper canines and premolars are considerably different from those of extant *Aotus*, in which the upper canines of both sexes are relatively small, showing very little sexual dimorphism. The new upper canine specimen of *A. dindensis*, despite a break in the tip, is evidently very robust and high, projecting well above the occlusal plane of the other teeth, and has a very deep, mesial, longitudinal groove, forming a heart-shaped occlusal outline (fig. 2). In P^2, as in the upper canine, the prominent paracone is much higher and sharper than the protocone. This reinforces earlier observations by Kay [10] on the P_2. In addition, distinct wear facets are observable at the distolingual face of the canine and at the mesial face of P^2, suggesting occlusal shearing between C^1-P^2 and P_2 (fig. 2). All these features indicate that this specimen is a male individual of a sexually dimorphic monkey. Compared to the canine specimens of extant male *Aotus* and *Saimiri*, the upper canine and P^2 of *A. dindensis* are actually more similar to those of *Saimiri*, which shows strong canine sexual dimorphism in both size and morphology (fig. 2).

The presence of canine sexual dimorphism in *A. dindensis* leads to some intriguing speculations regarding the ancestral state of the social systems and activity patterns of owl monkeys. Since the time of Darwin, many factors have been advocated to explain the evolution of canine sexual dimorphism: sexual selection from intermale

[1] Flynn et al. [9] obtained a slightly younger date for the Villavieja Formation. However, for this paper, the important point is that the new fossil specimen was found in a horizon that is stratigraphically higher than the type horizon.

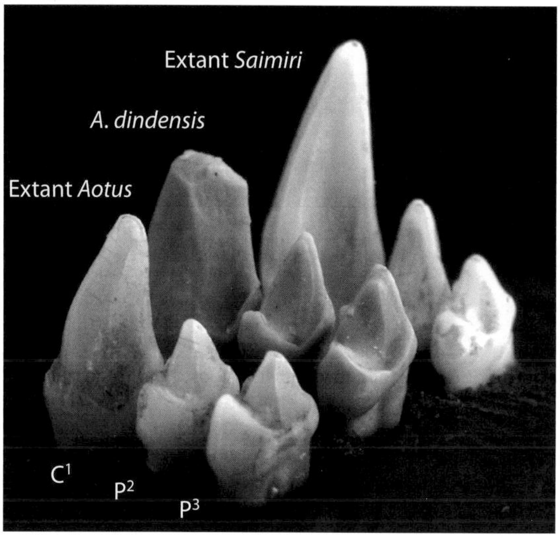

Fig. 2. Comparisons of upper canine and anterior premolars among extant/extinct *Aotus* and extant *Saimiri*. Distolingual views of left upper canine and P^{2-3} of extant male *Aotus* (front row), *A. dindensis* (middle row, epoxy casts of IGM-KU 98008–98010), and extant male *Saimiri* (tertial row). Upper canine and P^2 of *A. dindensis* appear to be much more similar to those of male *Saimiri* than to extant *Aotus*, though the apical tip of the upper canine specimen of *A. dindensis* is broken. The condition of upper canine and anterior premolars of *A. dindensis* definitely demonstrates this specimen is a male individual bearing the strong canine sexual dimorphism.

competition [11–15], predator defense [11, 16], body size allometry [17, 18], body-size sexual dimorphism [16], terrestriality [19], and phylogenetic inertia [20, 21]. According to recent comparative studies [12–14, 22], intermale competition is most strongly associated with canine dimorphism, although other factors such as body size, body-size dimorphism, and terrestriality are also partially associated with canine dimorphism. In *A. dindensis*, however, the effects of body size and terrestriality are unlikely to be the cause of the dimorphism because the dental size of *A. dindensis* is only slightly larger than of extant *Aotus*, and its arboreality has been suggested based on morphological analysis of the talus specimen (IGM-KU 8803), which is allocated to *A. dindensis* [23]. Although the influence of body-size dimorphism cannot be rejected in this discussion, even if it were the case, the presence of sexual dimorphism in *A. dindensis* indicates a non-monogamous social system of intense intermale competition.

Another interpretation should be considered for the presence of the large canines in *A. dindensis*. The canines could have been large in both males and females of this animal, as is true for extant gibbons. This hypothesis is improbable, however, for the following two reasons. First, considering the small, sexually monomorphic canines of extant *Aotus*, the existence of large, monomorphic canines in *A. dindensis* would require the reduction of canine size in both males and females, which is a less parsimonious explanation than the sexually dimorphic canine hypothesis. Second, sexual dimorphism has been reported in the canines of early anthropoids (such as proteopithecids, oligopithecids, parapithecids, and propliopithecids) from the late Eocene/early Oligocene of Fayum, Egypt [24, 25], and in the P_2 of the oldest fossilized platyrrhine, *Branisella*, from the late Oligocene of Salla, Bolivia [26]. The fossil evidence for canine sexual dimorphism in early, primitive anthropoids, and in fossil platyrrhines, suggest that canine sexual dimorphism is a primitive condition not only in the early platyrrhines but also in early anthropoids. Recent comparative analysis of extant primates also supports this hypothesis [2].

Among living anthropoid primates, however, the large canines are often displayed to competitors as well as used as actual weapons in intermale combat, because the escalation of combat with sharp, large canines may cause serious injury, disadvantaging both combatants [13]. Thus, the threat display of potential weapons is an important behavior in polygynous species with high intermale competition. The effective situation for such a display is not at night, but rather in the light of daytime. Hence, the retention of the large male canine in *A. dindensis* strongly suggests a diurnal activity pattern, which is consistent with the diurnal ancestry hypothesis deduced independently

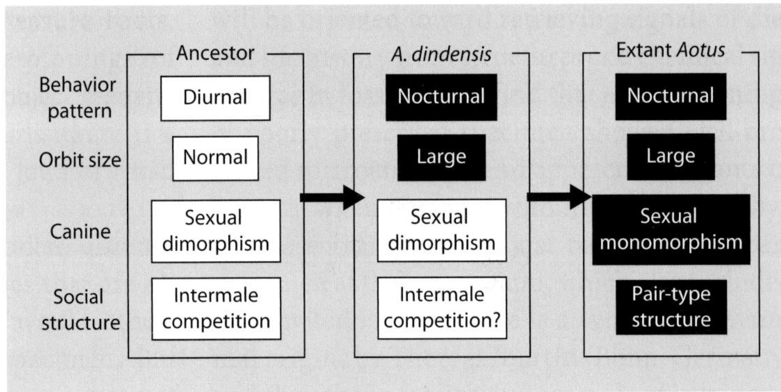

Fig. 3. Schematic diagram of evolution of nocturnality and reduction of sexual dimorphism in the *Aotus* lineage.

from anatomical evidence of retinal structure [3, 4] and from comparative analysis of extant primates. Recent intriguing behavioral and morphological reports on *Aotus azarae* of northern Argentine also strongly support this speculation: this extraordinary extant species exhibits a diurnal activity pattern and sexual dimorphism in the maxillary canines [27].

This logical conclusion inferred from fossil materials, however, appears as a curious contradiction to the morphological traits of *A. dindensis*: Why does the evidence for nocturnality (large orbits) and diurnality (canine sexual dimorphism) coexist in the middle Miocene *A. dindensis*? If *A. dindensis* had already adapted to nocturnal life as early as the middle Miocene, large male canines for threat display should be unnecessary in a nocturnal animal. In fact, maxillary canines are relatively small and blunt in both males and females of living owl monkeys. Considering the nocturnal activity pattern and monomorphic, small canines of extant *Aotus*, the most reasonable explanation for this contradiction is mosaic evolution of the craniodental morphology of owl monkeys. The large canine of *A. dindensis* may be a remnant feature of a diurnal, sexually dimorphic ancestor.

At the time of the ecological shift from diurnal to nocturnal habits, the diurnal ancestors of the owl monkeys had to quickly acquire large orbits to adapt to nocturnal life, as large orbits were probably critically important for the nocturnal vision of ancestral owl monkeys (fig. 3). On the other hand, the ancestral owl monkeys did not need to reduce 'useless' large male canines as quickly, but transformed them gradually to the small size seen in extant *Aotus*, as the retention of large male canines is not as critically disadvantageous as would be a lack of large orbits. From this perspective, it is interesting to note that the basic structure of the incisors, posterior premolars, and molars of owl monkeys has hardly changed during their evolutionary history (at least 15 Ma), most likely because these teeth were well adapted to their frugivorous and insectivorous diet.

The mosaic evolutionary process seen in the *Aotus* lineage suggests that adaptive metamorphosis may progress more quickly and smoothly than does the reductive transformation of useless features, and well-adapted features may hardly change in a relatively stable environment.

Acknowledgements

We thank all the staff of INGEOMINAS and Dr. Arnol Tovar E. of the Instituto Huilense de Cultura, Huila Department of Colombia, for their cooperation and support for our fieldwork in the Tatacoa Desert. We also thank Dr. Richard Kay of Duke University and Dr. Patricia Holroyd of the Museum of Paleontology, University of California, Berkeley, Calif., USA, for kindly showing us the fossil specimens of La Venta. For access to skeletal specimens of New World monkeys under their care, we thank Drs. Bruce Patterson and Bill Stanley of the Field Museum of Natural

History (Chicago) and Guy Musser and Wolfgang Fuchs of the American Museum of Natural History (New York). This work was supported by Overseas Scientific Research Funds (No. 10041165) and partly by the Global COE Program 'Formation of a Strategic Base for Biodiversity and Evolutionary Research: from Genome to Ecosystem' of the Ministry of Education, Culture, Sports, Science and Technology (MEXT), Japan (to Kyoto University).

References

1 Setoguchi T, Rosenberger AL: A fossil owl monkey from La Venta, Colombia. Nature 1987;326:692–694.
2 Heesy CP, Ross CF: Evolution of activity patterns and chromatic vision in primates: morphometrics, genetics and cladistics. J hum Evol 2001;40:111–149.
3 Ogden TE: The morphology of retinal neurons of the owl monkey *Aotes*. J Com Neur 1974;153:399–428.
4 Noback CE: The visual system of primates in phylogenetic studies; in Luckett P, Szalay FS (eds): Phylogeny of the Primates. New York, Plenum, 1975, pp 199–218.
5 Wright PC: The nocturnal primate niche in the New World. J Hum Evol 1989;18:635–658.
6 Takai M: New specimens of *Neosaimiri fieldsi* from La Venta, Colombia: a middle Miocene ancestor of the living squirrel monkeys. J Hum Evol 1994;27:329–360.
7 Takai M, Anaya F, Suzuki H, Shigehara N, Setoguchi T: A new platyrrhine from the middle Miocene of La Venta, Colombia, and the phyletic position of Callicebinae. Anthropol Sci 2001;109:289–307.
8 Takemura A, Takai M, Danhara T, Setoguchi T: Fission-track ages of the Villavieja Formation of the Miocene Honda Group in La Venta, Department of Huila, Colombia. Kyoto Univ Overseas Res Rep New World Monkeys 1992;8:19 27.
9 Flynn JJ, Guerrero J, Swisher GC: Geochronology of the Honda Group; in Kay RF, Madden RH, Cifelli RL, Flynn JJ (eds): Vertebrate Paleontology in the Neotropics: The Miocene Fauna of La Venta, Colombia. Washington, Smithsonian Institution Press, 1996, pp 44–59.
10 Kay RF: The physical relationships of extant and fossil Pitheciinae (Platyrrhini, Anthropoidea). J Hum Evol 1990;19:175–208.
11 Clutton-Brock TH, Harvey PH, Rudder B: Sexual dimorphism, socionomic sex ratio and body weight in primates. Nature 1977;269:797–800.
12 Kay RF, Plavcan JM, Glander KE, Wright PC: Sexual selection and canine dimorphism in New World monkeys. Am J Phys Anthropol 1988;77:385–397.
13 Plavcan JM, van Schaik CP: Intrasexual competition and canine dimorphism in anthropoid primates. Am J Phys Anthropol 1992;87:461–477.
14 Plavcan JM, van Schaik CP, Kappeler PM: Competition, coalitions and canine size in primates. J Hum Evol 1995;28:245–276.
15 Plavcan JM: Sexual dimorphism in primate evolution. Yearb Phys Anthropol 2001;44:25–53.
16 Leutenegger W, Kelly JT: Relationship of sexual dimorphism in canine size and body size to social, behavioral, and ecological correlates in anthropoid primates. Primates 1977;18:117–136.
17 Leutenegger W, Cheverud J: Correlates of sexual dimorphism in primates: ecological and size variables. Int J Primatol 1982;3:387–402.
18 Leutenegger W, Cheverud J: Sexual dimorphism in primates; in Jungers WL (ed): Size and Scaling in Primates. New York, Plenum, 1985, pp 33–50.
19 Harvey PH, Kavanagh M, Clutton-Brock TH: Sexual dimorphism in primate teeth. J Zool, London 1978;186:475–485.
20 Cheverud JM, Dow MM, Leutenegger W: The quantitative assessment of phylogenetic constraints in comparative analyses: sexual dimorphism in body weight among primates. Evolution 1985;39:1335–1351.
21 Lucas PW, Corlett RT, Luke DA: Sexual dimorphism of tooth size in anthropoids. Hum Evol 1986;1:23–39.
22 Plavcan JM: Reconstructing social behavior from dimorphism in the fossil record; in Plavcan JM, Kay RF, Jungers WL (eds): Reconstructing Behavior in the Primate Fossil Record. New York, Kluwer Academic/Plenum, 2002, pp 297–338.
23 Gebo DL, Dagosto M, Rosenberger AL, Setoguchi T: New platyrrhine tali from La Venta, Colombia. J Hum Evol 1990;19:737–746.
24 Fleagle JG, Kay RF, Simons EL: Sexual dimorphism in early anthropoids. Nature 1980;287:328–330.
25 Simons EL, Plavcan JM, Fleagle JG: Canine sexual dimorphism in Egyptian Eocene anthropoid primates: *Catopithecus* and *Proteopithecus*. Proc Natl Acad Sci USA 1999;96:2559–2562.
26 Takai M, Anaya F, Shigehara N, Setoguchi T: New fossil materials of the earliest New World monkey, *Branisella boliviana*, and the problem of platyrrhine origins. Am J Phys Anthropol 2000;111:263–281.
27 Fernandez-Duque E, Bravo SP: Population genetics and conservation of owl monkeys (*Aotus azarai*) in Argentina: a promising field site. Neotropical Primates 1997;5:48–50.

Dr. Masanaru Takai
Primate Research Institute, Kyoto University
41 Kanrin, Inuyama
Inuyama 484-8506 (Japan)
Tel. +81 568 63 0533, Fax +81 568 63 0536, E-Mail takai@pri.kyoto-u.ac.jp

Koppe T, Meyer G, Alt KW (eds): Comparative Dental Morphology.
Front Oral Biol. Basel, Karger, 2009, vol 13, pp 60–64

Observation of Lateral Mandibular Protuberance in Taiwan macaque *(Macaca cyclopis)* Using Computed Tomography Imaging

Shintaro Kondo[a] · Munetaka Naitoh[b] · Chiharu Futagami[a] · Hajime Hanamura[a] · Kenichi Goto[b] · Eiichiro Ariji[b] · Masanaru Takai[c]

Departments of [a]Anatomy and [b]Oral and Maxillofacial Radiology, School of Dentistry, Aichi-Gakuin University, Nagoya, and [c]Primate Research Institute, Kyoto University, Inuyama, Japan

Abstract

Morphological characteristics of the protuberance on the external surface of the mandible in Taiwan macaque (*Macaca cyclopis*) was investigated using cone-beam computed tomography. We observed 49 skulls of *M. cyclopis*. Of 7 skulls with deciduous and mixed dentitions in which M2s did not erupt, the protuberance was not found. Of the 13 skulls with mixed and permanent dentitions in which M2s had erupted, a palpable protuberance was found in one specimen. Of the 29 samples in which M3s had erupted completely, a perceptible protuberance was found in 2 samples, and palpable protuberance was found in 8 samples. Thus, the protuberance was found in 10 samples of the 29 samples with complete dentitions (34.5%), and the emergence of the protuberance may have been related to mandibular growth. In the case of the well-developed protuberance, it extended from the P4 to M3 region but did not extend to the mental foramina. By using cone-beam computed tomography, it was determined that the protuberance was composed of cortical bone and was the thickest in M2 region. Since the protuberance consisted of homogeneous cortical bone, it was considered to be the result of normal bone growth similar to the mandibular torus in humans.

We found a mandible with a well-developed protuberance on its external surface, during our survey on the skulls of the Taiwan macaque (*Macaca cyclopis*) at the Primate Research Institute, Kyoto University (Inuyama, Japan). The protuberance extended from the premolar to the molar region, and its volume was almost equal to that of a normal mandibular corpus (fig. 1). There have been no reports on such protuberances on macaque mandibles, as far as we know. Our general aim in this study is to describe this protuberance morphologically, to observe the internal structure of it using cone-beam computed tomography (so that we could compare the protuberance and mandibular corpus), and to present a brief discussion of the possible cause of the protuberance while making comparisons with similar structures in humans.

Materials and Methods

Forty-nine skulls of *M. cyclopis* (housed in the Primate Research Institute, Kyoto University) were initially

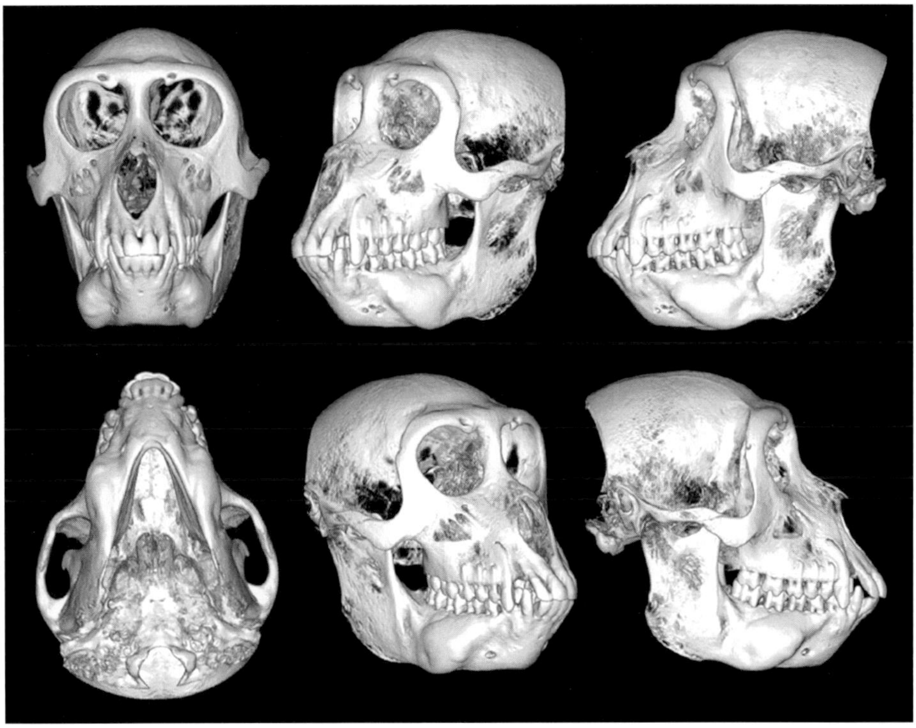

Fig. 1. Three-dimensional images reconstructed from cross-sectional tomography images (KUPRI #6689, female).

examined. The external structure of the mandible was investigated using palpation. For 4 mandibles which represented the range of typical external structure, the internal structure was investigated using cone-beam computed tomography (Asahi Roentgen Ind., Co., Kyoto, Japan). The voxel size was 0.2 × 0.2 ×0.2 mm. The scan was set at 80 kV and 5 mA, as recommended by the manufacturer. The DICOM files of the axial images were saved to a portable hard disk. Two-dimensional images of various planes and three-dimensional images were reconstructed on a personal computer using three-dimensional visualization and measurement software (OsiriX, ver2.7.5) [1].

Results

Morphological Description of the Well-Developed Protuberance

The specimen that we mentioned above was a female adult in which the M3s were fully erupted (KUPRI #6689). Three-dimensional images from cross-sectional tomography were shown in figure 1. The protuberance was found bilaterally, in the P3 to M3 region, but it did not extend downward to the mental foramina. The surface was not rugged, and fairly smooth. Cone-beam computed tomography showed that the protuberance was composed of cortical bone and was thickest in the M2 region (fig. 2), where it was as thick as the mandibular corpus itself. The cortical bone was high density and homogeneous, but some canals were also visible.

Does Such a Protuberance Exist in the Other Individuals?

We observed 49 individuals of *M. cyclopis* in order to confirm whether such a protuberance appeared only in the one specimen or not. The protuberance showed variable expressions

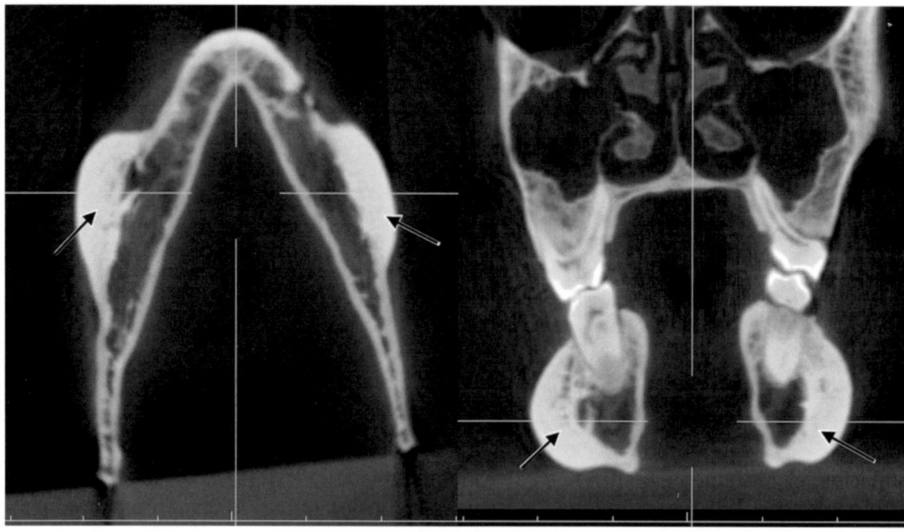

Fig. 2. Reconstructed two-dimensional images from cross-sectional tomography images. Horizontal (left) and frontal (right) images in M2s region (KUPRI #6689, female). Arrows indicate the protuberance (cortical bone).

morphologically. Three degrees of expressions of the protuberance are shown in figure 3: (a) bilateral protuberances were slight and only noticeable by palpation, (b) on one side a protuberance was only recognized by palpation, but on the other side it was more pronounced, and (c) a well-developed protuberance was present bilaterally.

Of the 29 samples in which M3s had erupted completely, a well-developed protuberance was visible in only 2 samples. One of these was found unilaterally. However, a weak protuberance, only noticeable by palpation, was found in 8 samples. Thus a protuberance was found in 10 of 29 samples (34.5%). A protuberance was found in 3 of 12 males (25.0%), and 7 of 17 females (41.2%). Although the protuberance was more frequently found in females than in males, the difference were not statistically significant. Thus, the protuberance was a fairly common structure in *M. cyclopis*. In many cases, the protuberance extended from the P4 to M3 region.

Interestingly, none of the 7 samples of deciduous and mixed dentitions in which M2s had not erupted exhibited the protuberance. Of the 13 samples of mixed and permanent dentitions in which M2s *had* erupted, a palpable protuberance was found in only one specimen. Thus the protuberance didn't seem to appear until the M2s had erupted. Its emergence may thus be, in some way, related to the mandibular growth.

Relationship between the Protuberance and Cortical Bone Thickness

We measured the thickness of the buccal cortical bone and that of the mandibular corpus in the M2 region on the frontal images reconstructed from the CT scans. Thus four samples (2 males and 2 females) were measured; 3 had various protuberances, and the other sample had no discernible protuberance.

The thickness of the mandibular corpus except for the buccal cortical bone thickness ranged

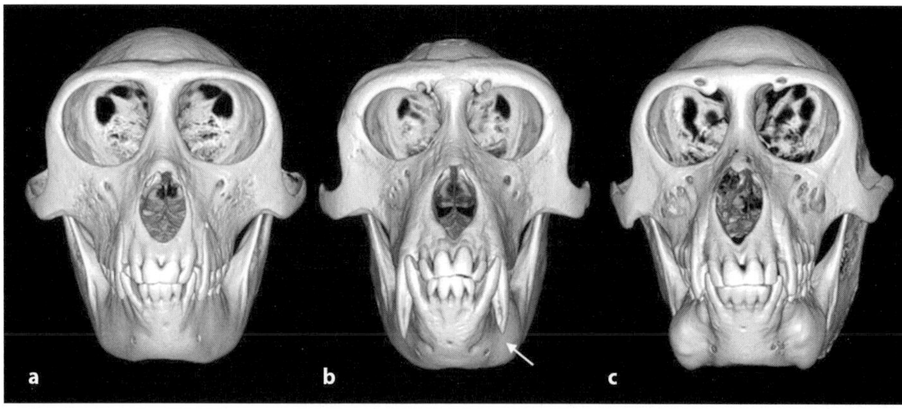

Fig. 3. Three types of the protuberances were shown (Three dimensional images, frontal view). **a** Palpable protuberance, bilateral (KUPRI #1511, female). **b** Palpable (right side) and perceptible (left side, arrow) protuberance (KUPRI #1358, male). **c** Well-developed protuberance, bilateral (KUPRI #6689, female).

from 7.9 to 9.0 mm. The buccal cortical bone thickness ranged from 1.5 to 6.3 mm, with the largest value found in a large protuberance (fig. 1, 2, 3c). In the case of palpable protuberances, the buccal cortical bone was not thickened (1.5 mm). Thus the thickness of the cortical bone did not always accord with the macroscopic observation of the protuberance.

Discussion

Since the protuberance in *M. cyclopis* consisted of homogeneous cortical bone without trabeculae, it was considered the result of normal bone growth. The mandibular torus in humans is a bony exostosis of varying size on the medial side of the mandible, and usually bilateral [2]. It is most commonly found in the vicinity of the second premolar to first molar, and it usually has a smooth surface [2]. The torus is normally composed of cortical bone with homogeneous mineralization, but it may also contain a marrow space and trabeculae [3]. The protuberance in macaques is thus similar to the torus in humans morphologically, even though it is found on the lateral side of the mandible in macaques and the medial side in humans.

We have suggested that the protuberance in macaques does not appear until the M2s are erupted completely. Igarashi et al. [4] showed that the mandibular torus in humans developed in correlation with the degree of dental attrition, the number of teeth and age. They suggested that mandibular tori were promoted by bony response to masticatory stress in normal chewing and other factors correlated with age. Their results are similar to ours for the macaque's protuberance, and they parallel the findings of Eggen and Natvig [5], who showed that the incidence of mandibular tori in humans was correlated with the number of teeth – to the point of actually *decreasing* as the number of teeth decreased with age.

Obviously, the documentation of actual stresses and strains in mastication is not easy to do. Work with nonhuman primates has shown that thickening of the mandibular corpus in the molar region is an effective adaptation to increased torsion about the long axis of the working-side mandibular corpus during unilateral mastication

[6]. However, an alternative cause of mandibular thickening might be that large local stress concentrations on the post-canine teeth could cause direct shear of the corpus that in turn might stimulate cortical bone growth.

Of course, stress-strain interactions like these do not occur in a biological vacuum. Thus, for instance, since mandibular tori appeared in a fairly large number of Aleuts, Morrees et al. [7] suggested that there was a genetic basis for mandibular tori in humans. Johnson et al. [8] reiterated this in a study of differences in the familial incidence of the trait. More recently, Jainkittivong and Langlais [9] showed that the etiology of exostoses (including mandibular tori) involved the complex, multifactorial interplay of genetic and environmental factors. Given the potential complexity of the growth and development of the lateral mandibular protuberance, and our small sample sizes, this is probably the best explanation, for now, of the results of this study on the Taiwan macaque (*M. cyclopis*).

Acknowledgment

This study was supported by a Grant-in-Aid for Scientific Research (C) from the Ministry of Education, Culture, Sports, Science and Technology, Japan (No. 20570228), and the Cooperation Research Program of Primate Research Institute, Kyoto University.

References

1 Rosset AR, Spadola L, Ratib O: Osirix: an open-source software for navigating in multidimensional DICOM images. J Digital Imaging 2004;17:205–216.
2 Sellevold BJ: Mandibular torus morphology. Am J Phys Anthropol 1980;53:569–572.
3 Yonetsu K, Nakamura T: CT of calcifying jaw bone diseases. Am J Roentogenol 2001;177:937–943.
4 Igarashi Y, Ohzeki S, Uesu K, Nakabayashi T, Kanazawa E: Frequency of mandibular tori in the present-day Japanese. Anthropol Sci 2008;116:17–32.
5 Eggen S, Natvig B: Relationship between torus mandibularis and number of present teeth. Scand J Dent Res 1986;94:233–240.
6 Hylander WL: The functional significance of primate mandibular form. J Morphol 1979;160:223–240.
7 Moorrees CFA, Osborne RH, Wilde E: Torus mandibularis: its occurrence in Aleut children and its genetic determinants. Am J Phys Anthropol 1952;10:319–330.
8 Johnson CC, Gorlin RJ, Anderson VE: Torus mandibularis: a genetic study. Am J Hum Gentics 1965;17:433–442.
9 Jainkittivong A, Langlais RP: Buccal and palatal exostoses: prevalence and concurrence with tori. Oral Surg Oral Med Oral Pathol Oral Radiol Endod 2000;90:48–53.

Shintaro Kondo
Associate Professor
Department of Anatomy, School of Dentistry, Aichi-Gakuin University
1–100 Kusumoto-cho, Chikusa-ku
Nagoya 464-8650 (Japan)
Tel. +81 527512561, Fax +81 527525988, E-Mail kondos@dpc.agu.ac.jp

Koppe T, Meyer G, Alt KW (eds): Comparative Dental Morphology.
Front Oral Biol. Basel, Karger, 2009, vol 13, pp 65–67

Dental Tissues: An Introduction

M. Christopher Dean

Cell and Developmental Biology, University College London, London, UK

In his introduction to the published volume that resulted from the first International Symposium on Dental Morphology that took place at Fredensborg, Denmark, in September 1965, Albert Dahlberg wrote that material pertinent to modern mammalian dentitions was scattered in journals and texts of many disciplines and in many languages. This circumstance, he noted, had led to some difficulty in communication and synthesis between ideas and workers. Following the success of the two meetings organised by The Society for The Study of Human Biology, especially the one in London in 1962, which had resulted in *Dental Anthropology* the edited volume by Don Brothwell in 1963, Dahlberg, Pedersen and Alexandersen formulated a plan to bring together scientists at a 3-day international meeting with a wide scope. The plan was to spend a day discussing phylogenetic aspects of the dentition, a day on ontogenetic issues and a day on dental morphology and genetics. At this first meeting 65 participants including palaeontologists, embryologists, anatomists, zoologists, geneticists and odontologists first circulated reports and previously published material on aspects of the dentition ahead of the meeting. These were then formally presented and discussed. Twenty-five original papers that resulted from this meeting were then published in the *Journal of Dental Research*, volume 46, number 5, 1967. Even today, this volume remains an impressive interdisciplinary collection of research papers that is well worth revisiting.

The same spirit of synthesis and communication across diverse areas of dental biology underpins the success of these meetings. At the most recent International Symposium on Dental Morphology, the 14th held in Greifswald, Germany, during August 2008, the number of sessions that each now define an area of dental morphology had risen to 6, one more than at the previous 13th meeting in Łodź, Poland, in August 2005. How, one might ask, has 'dental tissues' come to win a place of its own in this latest symposium on dental morphology? Quite simply, each of the papers published in this section do more than just describe dental tissues; they each aim to inform or explain some aspect of adult dento-alveolar morphology. Adult forms do not evolve or transform into other adult forms. Embryonic, developmental and growth processes constantly and randomly generate variable adult morphologies and pathologies through a very complex sequence of events. Dental tissues are special in that they preserve a record of their own development that can explain adult morphologies by revealing some of the processes that underlie them. This becomes increasingly powerful when genetics and molecular biology can be linked to adult morphologies and pathologies. The degree to which tooth tissues and bone are modular or independent in their development and growth is also an important issue for dental morphologists. Studies at the tissue level can reveal either gradual transitions of tissue types or unlikely associations between the many components of teeth and jaws that are explained through an understanding of gene expression.

In the first paper of this section, Dean uses incremental markings in enamel and dentine to show how variation in growth rates might explain how tooth crowns can grow either taller or shorter in the same period of time. Initial fast rates of differentiation along the enamel dentine junction contribute most to tall crowns, but slower rates of growth over a longer period of time in the lateral and cervical regions can prolong total crown formation times in shorter crowns. Root growth rates along the cement dentine junction also vary and this study demonstrates some of the differences that existed in root growth rates among fossil hominin teeth that can still be retrieved using incremental lines in dentine.

Diekwisch and coworkers make an important link between the morphology of mature enamel and the molecular composition and structure of enamel proteins. Enamelin is the predominant enamel protein in shark enameloid, whereas in reptiles, amphibians and mammals amelogenins predominate. Amelogenins control the rate and direction of growth of the hydoxyapatite crystals in enamel. They make up 90% of all mammalian enamel proteins. Aprismatic enamel predominates in the majority of squamates and amphibians but the characteristically complex prismatic enamel microstructure of mammals might depend upon the physico-chemical properties of the amelogenin molecule and in turn upon certain unique amino acid sequences found only in mammalian amelogenins. Diekwisch and coworkers identified four new amelogenin amino acid sequences from three amphibian species and compared these non-mammalian amelogenins with those known for mammals. The authors report that amphibian amelogenin proteins are shorter, contain less proline and glutamine and most significantly, have shorter polyproline-tripeptide repeat stretches than their mammalian counterparts. The next step is to clinch the link between increased rigidity of mammalian amelogenins and the orderly growth of longer parallel apatite crystals typical of mammalian prismatic enamel.

Kieser's group describe the dento-alveolar microanatomy of *Sphenodon,* the Tuatara, known only today from islands off the coast of New Zealand. Both SEM and light microscopy reveal that well-differentiated tissues exist in their teeth and jaws, contrary to some previous reports. A thin layer of prismless enamel covers the crown. Dentine in these teeth grades from tubular in the cuspal region, with many secondary branches, to a more sclerosed tissue cervically that increasingly contains cell inclusions towards the basal bone of attachment. The basal dentary bone contains waves of cell inclusions that might represent alternating fast and slow bouts of bone deposition. The continuous microanatomical gradient from dentine to bone raises interesting questions about the developmental origins and definition of tissue types. Equally interesting is the evidence of remodelling within the lamella-like bone of attachment to which the tooth is ankylosed.

Smith and Reid present evidence that the ridges often visible on tooth root surfaces (periradicular bands) are associated with the long-period incremental markings within the root dentine. This study provides yet more evidence that human, and other primate teeth, express a periodic slowing down and speeding up of enamel and dentine formation that is greater than a day. In this study, the counts of incremental markings within the dentine and of periradicular bands were the same between defined markings. The fact that in both enamel and dentine these long-period rhythms are visible on the tooth surface provides an important accessible temporal record of tooth growth that can be used to calculate rates of crown, and in this case root growth, as well as tooth root formation times. The next challenge will be to reconcile this new evidence with other data that suggest enamel perikymata may not always express the same periodicity as periradicular bands on tooth roots of the same individual.

Lesot and coworkers describe the effects of a mutation of the *Eda* gene on developing teeth and bone. This gene encodes for ectodysplasin-A1. The mutation expresses itself as a severe X-linked hypohydrotic ectodermal dysplasia (XLHED), which is often associated with oligodontia. The effects of the mutation were investigated using medical CT scans and bone densitometric profiles in patients being assessed for treatment. Changes to the structure and morphology of the bone, which included a marked increase in bone density, appear not to be secondary to the dental pathology, but rather, are a direct consequence of the gene mutation acting on bone-forming cells. Furthermore, these patients showed an altered pattern of skeletal growth and there is some suggestion that bones formed in membrane may be affected to a greater degree than those preformed in cartilage. This study illustrates the diverse skeletodental pathology that a single gene mutation can bring about but importantly, how many of these can be explained by a knowledge of which cell types normally express Eda.

M. Christopher Dean, PhD
Department of Cell and Developmental Biology, University College London
Gower Street
London, WC1E 6BT (UK)
Tel. +44 20 7679 0462, E-Mail chris.dean@ucl.ac.uk

Koppe T, Meyer G, Alt KW (eds): Comparative Dental Morphology.
Front Oral Biol. Basel, Karger, 2009, vol 13, pp 68–73

Extension Rates and Growth in Tooth Height of Modern Human and Fossil Hominin Canines and Molars

M. Christopher Dean

Cell and Developmental Biology, University College London, London, UK

Abstract

The aim of this study was to describe similarities and differences in the way modern and fossil hominin teeth grow in height. Measurements from longitudinal ground sections of 7 modern human canines and 19 first permanent molars were used to calculate extension rates in the crowns and roots and to plot distance curves for growth in tooth height. These were compared with identical data for 3 fossil hominin teeth attributed respectively to *Paranthropus robustus*, *Homo erectus* and *Homo neanderthalensis*. Enamel extension rates in each of the three fossil taxa fell within the range of modern humans. Root extension rates in the fossil taxa also fell within modern human ranges but differed in their pattern with either an early or late marked increase in root length. Extension rates in the canine crowns were higher in cuspal enamel than in lateral enamel. Combinations of high or low cuspal enamel extension rates, with either longer or shorter times taken to form lateral enamel, explain how crown formation times may vary independently of completed crown heights.

Longitudinal records of tooth growth exist that are embedded within the incremental structure of enamel and dentine. These can be retrieved, even from fossil teeth, to reconstruct growth that occurred in tooth germs. To date, more studies have used incremental markings in enamel and dentine to determine the timing of crown and root formation than have used them to reconstruct the growth processes and/or different patterns of growth that occurred during odontogenesis. In this study, three questions are addressed. First, do rates of extension (i.e. the rates of terminal differentiation of odontoblasts and ameloblasts as defined by Shellis [1]) along the crowns of modern human and fossil teeth differ from one another? The second question is whether differences in extension rates can explain how some long teeth take a short time to form their crowns and how some short teeth take a long time to form their crowns? To answer this question, the extension rates in a comparatively short (10.7 mm) robust australopith fossil canine crown with a short crown formation time (3.2 years) was compared with a sample of modern human canines of widely varying heights and crown formation times. The last question addressed here is whether the crown and root extension rates of two fossil hominins available for study (a Neanderthal first permanent molar and a *Homo erectus* first permanent molar) fall within or beyond those of modern humans and do any obvious differences exist between these fossil molars?

Materials and Methods

Ground sections of 7 modern human canines and 19 first permanent molars of diverse geographical origin were chosen that showed both clearly visible daily cross striations expressed close to the EDJ in the crown, as well as clearly visible accentuated or regular long-period markings expressed along the CDJ as far along the root as possible. For each section, counts of the number of daily cross striations were made within a 200-µm distance from the EDJ in the mid-occlusal, mid-lateral and mid-cervical regions of the crown. Across the whole modern human sample this cross striation count varied between 68 and 92 days with a mean value of 80. However, within an individual tooth section, the number was consistent in all regions to within 5 days. Therefore, for each section the average number of days to form 200 µm of enamel from the EDJ was estimated on this basis. Following methods described previously [1–4], repeated calculations of length in micrometers along the EDJ, and of the number of days taken to form that length, were cumulated for each 200-µm formation period. The first calculation was at the dentine horn and the last at the enamel cervix. In each case the number of days formation defined by the methods was always 200, except for the last fraction along the EDJ remaining at the cervix.

A similar approach was used to that previously described [5–7] to calculate the same variables in the tooth roots. This was done by beginning at the enamel cervix along the CDJ and continuing as far into the tooth root as possible. A range of daily dentine formation rates just deep to the granular layer of Tomes (GLT) was calculated in the following way. Ten calculations of root extension rates were first made from previously published data [8] where the time in days between tetracycline labels had been calibrated using daily enamel cross-striations. For two human tooth roots in this sample (across lines 14–23 in Dean et al. [8]) an average of 10 calculations were made of extension rates at different positions along each tooth root. While the extension rates differed at different points along the root, the back-calculation of the daily rates of dentine formation were consistent in each tooth root. For a 200-µm length along a dentine tubule beginning just deep to the GLT, average daily rates of dentine formation varied between 2.0 µm/day in a lower central incisor and 2.5 µm/day in a lower first permanent molar.

Previously [5–7], a rate of 2.5 µm/day has been used to calculate root extension rates in modern humans but Smith et al. [9] have argued daily rates closer to 2.0 µm/day are more appropriate for some teeth based on data collected for *Pan*. For *Pan* this does indeed seem reasonable. While this potential small difference between *Pan* and *Homo* and between different modern human tooth types remains unresolved, the slower rate of 2.0 µm/day determined here for the lower central incisor was used for all the modern human teeth in this study, such that a 200-µm distance along a dentine tubule just deep to the GLT was estimated to take 100 days to form. Again, following previously described methods [1, 2, 4] repeated calculations of length in micrometers along the CDJ and of the number of days taken to form that length were cumulated for each corresponding 200-µm formation period along a dentine tubule. The initial calculation was at the enamel cervix and the last was as far along the CDJ as possible to a point where accentuated or long-period markings became indistinct.

Data previously collected and published from three ground sections of fossil hominin teeth were then compared with the data collected for modern humans. SK-63 is a lower canine from the site of Swartkrans, South Africa and attributed to *Paranthropus robustus* [10]. S7–37 is an upper first permanent molar attributed to *Homo erectus* from Sangiran, Java [11] and BD-J4-C9 is a lower left first permanent molar attributed to a Neanderthal from Abri Bourgeois-Delaunay at the site of La Chaise-de-Vouthon, Charente, France [7]. For each of these fossils the number of enamel cross striations over a 200-µm prism length had previously been counted directly (SK-63 = 62 days; BD-J4-C9 = 66 days and S7–37 = 60 days). In one of the fossils (S7–37) daily incremental lines in the dentine could also be seen at various points along the root. Therefore, in this specimen these direct observations were used to calculate a formation rate, which was 2.5 µm/day for a 200-µm length along a dentine tubule just deep to the GLT. One fossil tooth (SK-63) had little or no root dentine formed and so no data for root growth could be presented. In the case of BD-J4-C9 no daily increments were visible in dentine so values of both 2.0 and 2.5 µm/day were used and plotted separately for comparison.

Results

Figure 1 shows the increase in length along the EDJ in the crown of SK-63 (solid circles) against time. These data are set against the modern human data for canine enamel extension rates along the EDJ, which were then continued along the CDJ as root extension rates. The data for the *Paranthropus robustus* canine fall comfortably within the range for the modern human teeth and give no hint that rates of extension along

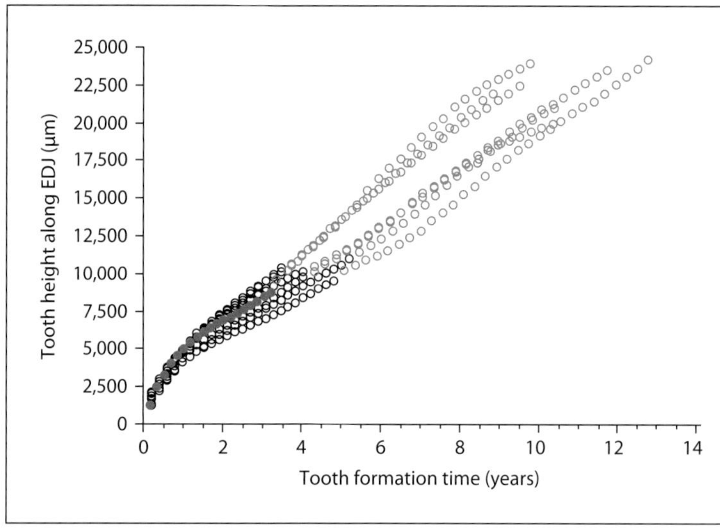

Fig. 1. Human canines and SK-63. Distance curve of tooth height (μm) along the EDJ against time (years). The data for modern human canines depict crown growth in bold open circles and root growth in faint open circles. The data for growth along the EDJ in SK-63 appear as filled-in grey circles.

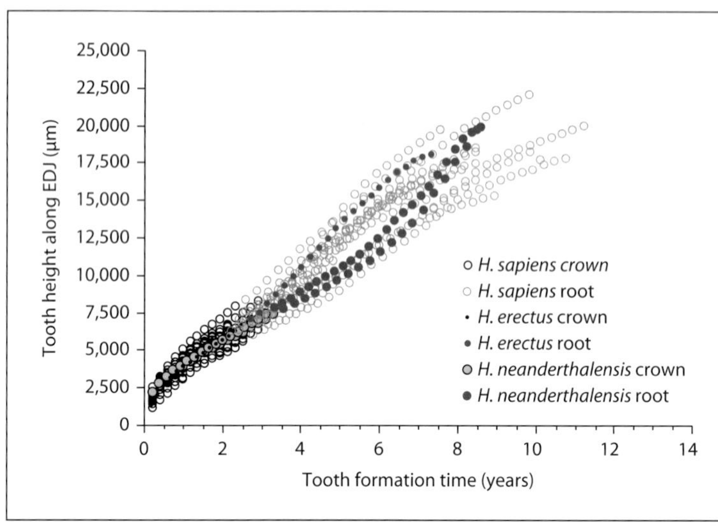

Fig. 2. *H. sapiens, H. erectus* and Neanderthal M1. Distance curve of tooth height (μm) along the EDJ against time (years). The data for modern human first permanent molars depict crown growth in bold open circles and root growth in faint open circles. The crown data for the *H. erectus* molar are shown with small in-filled light-grey circles and the root data as small in-filled dark-grey circles. The crown and root data for the Neanderthal molar are both depicted as large in-filled grey circles. Two plots for the Neanderthal root are shown, one using 2.0 μm/day and the other using 2.5 μm/day for daily rates of root dentine formation.

the EDJ are faster than in modern humans for this tooth type, especially so towards the cervix. Figure 2 shows the extension rate data for modern human first permanent molars together with those for the *H, erectus* and *H. neanderthalensis* first permanent molars. Again, the extension rate along the EDJ in the fossil specimens falls comfortably within the modern human range. Similarly, in the roots, neither fossil falls outside the range of the human data although there is an

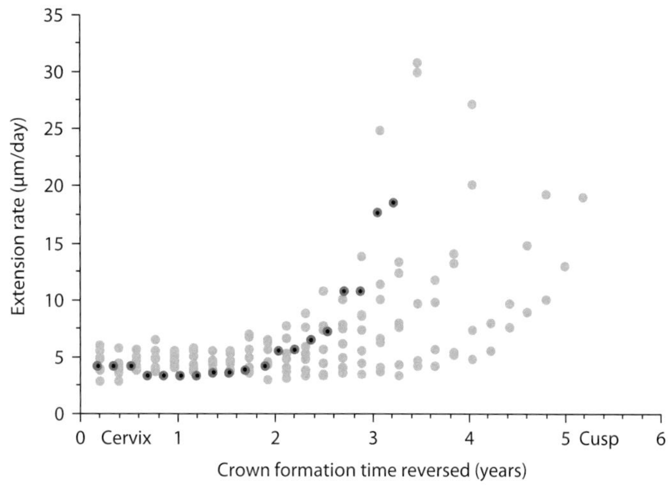

Fig. 3. Human canines and SK-63. Extension rates (µm/day) along the EDJ are plotted against time (years) as if each canine tooth grew from the cervix to the dentine horn. This reveals that some canines have high initial extension rates in the cusp (~30 µm/day) and others lower rates (~20 µm/day). There are also differences in the time taken to form lateral enamel where rates usually fall in the range 3–6 µm/day.

earlier rise, or spurt, in height along the CDJ in the root of the *H. erectus* root, but a later rise, or spurt, in height along the CDJ of the Neanderthal molar root. The two plots for the Neanderthal molar root demonstrate the degree of difference that results from using either 2.0 or 2.5 µm/day in these calculations. In no sense does this appear to alter the picture that distinguishes the pattern of Neanderthal root growth from that in modern humans or *H. erectus*.

Figure 3 shows the range of maximum crown heights and the range of times taken to form canine crowns in the modern human sample together with the *Paranthropus robustus* canine crown data. In this plot, the data along the X-axis have been reversed to imply growth of the crown begins at the cervix and ends at the dentine horn. This pulls longer and shorter crown formation times apart in a way that is more easily visualised than if the data were plotted from cusp to cervix. None of the extension rate estimates along the EDJ fall outside those of the modern human sample.

Discussion

Many studies have used surface perikymata on enamel to estimate the time taken to form enamel in modern human and fossil teeth [12–15]. This is justified when the periodicity (the number of days between adjacent long-period incremental markings and perikymata) and the time taken to form cuspal enamel can be estimated with some confidence. However, the results of this study suggest it may be misleading to go further than this and use the spacing of perikymata on the surface of a tooth to estimate changing rates of extension in the crown. In the first place, anterior teeth only express ~50% of the EDJ length in the form long-period striae of Retzius that project towards and crop out on the tooth surface as perikymata. Posterior teeth with thick cuspal enamel express much less than 50% of EDJ length on the enamel surface. In the second place, the distribution of enamel thickness within the crown and the contour of the crown shape itself underlie differences in perikymata spacing [16]. The perikymata spacings on the *P. robustus* (SK-63) enamel surface remain wide, even towards the cervix, compared with the majority of modern human teeth where they become more tightly packed together. However, the extension rates calculated in this study do not reflect this pattern, indeed they are comparatively low in both cuspal and lateral enamel. The implication is that widely spaced perikymata at the cervix of some fossil teeth may result from faster-formed daily increments of enamel maintaining a wider space between adjacent cervical perikymata rather than from faster cervical extension rates. This hypothesis, however, is yet to be tested but might also explain the range of variation in perikymata packing patterns among geographically diverse populations of modern humans today and even perhaps some of the variation that existed among Neanderthals and their contemporary Upper Palaeolithic human neighbours [13–15].

Reconstructing growth in length along the EDJ and the CDJ reveals how fossil teeth grew their crowns and roots in a similar way to modern human teeth, but it also highlights how the *H. erectus* and Neanderthal molar tooth roots fall at opposite extremes of the modern human range (fig. 2). How this might relate to the process of and the timing of tooth eruption is both intriguing and interesting – and a topic for future research [5, 6].

The data for enamel extension rates in the modern human canine sample and the robust australopith fossil canine presented in this study (fig. 3) also explain how longer or shorter teeth can grow their crowns either fast or slowly. They show why tooth crown height is not directly related to crown formation time. Fast extension rates in the cusp, combined with short periods of time spent forming lateral and cervical enamel along the EDJ, would result in tall canines taking the least amount of time to form their crowns. Slower extension rates in the cusp combined with longer periods of time spent growing lateral and cervical enamel would result in shorter canines taking the longest time to form their crowns. Clearly, all combinations are possible. The *Paranthropus robustus* canine, SK-63, is short and only 10.7 mm long with a comparatively short crown formation time in human terms of 3.2 years [10]. It achieved this height with relatively slow values for extension rates in the cusp combined with a relatively short period of lateral enamel formation.

Acknowledgements

I thank Thomas Koppe for inviting me to give a keynote talk at the 14th International Symposium on Dental Morphology in Greifswald, August 2008, and The Leverhulme Trust for funding a project to study rates of dentine formation in primates. I am grateful for the helpful comments of three referees.

References

1 Shellis RP: Variations in growth of the enamel crown in human teeth and a possible relationship between growth and enamel structure. Archs Oral Biol 1984;29:697–705.
2 Risnes S: Enamel apposition rates and the prism periodicity in human teeth. Scand J Dent Res 1986;94:394–404.
3 Risnes S: Growth tracks in dental enamel. J Hum Evol 1998;35:331–350.
4 Dean MC: A comparative study of cross striation spacings in cuspal enamel and of four methods of estimating the time taken to grow molar cuspal enamel in *Pan, Pongo* and *Homo*. J Hum Evol 1998;35:449–462.
5 Dean MC: A radiographic and histological study of lower first permanent molar root growth during the supraosseous eruptive phase. J Hum Evol 2007;53:635–646.
6 Dean MC, Vesey P: Preliminary observations on increasing root length during the eruptive phase of tooth development in modern humans and great apes. J Hum Evol 2008;54:258–271.
7 Macchiarelli R, Bondioli L, Debénath, A, Mazurier A, Tournepiche J-F, Birch W, Dean C: How Neanderthal molar teeth grew. Nature 2006;444:748–751.

8 Dean MC, Beynon AD, Reid DJ, Whittaker DK: Longitudinal study of tooth growth in a single individual based on long and short period incremental markings in dentine and enamel. Int J Osteoarchaeol 1993;3:249–264.
9 Smith TM, Reid DJ, Dean MC, Olejniczak AJ, Martin LB: Molar development in common chimpanzees (*Pan troglodytes*). J Hum Evol 2007;52:201–216.
10 Dean MC, Beynon AD, Thackeray JF, Macho G: Histological reconstruction of dental development and age at death of a juvenile *Paranthropus robustus* specimen, SK 63, from Swartkrans, South Africa. Am J Phys Anthropol 1993;91:401–419.
11 Dean C, Leakey MG, Reid D, Schrenk F, Schwartz GT, Stringer C, Walker A: Growth processes in teeth distinguish modern humans from *Homo erectus* and earlier hominins. Nature 2001;414:628–631.
12 Bromage TG, Dean MC: Re-evaluation of the age at death of immature fossil hominids. Nature 1985;317:525–528.
13 Ramirez-Rozzi FV, Bermudez de Castro JM: Surprisingly rapid growth in Neanderthals. Nature 2004;428:936–939.
14 Guatelli-Steinberg D, Reid DJ, Bishop TA: Anterior tooth growth periods in Neanderthals were comparable to those in modern humans. Proc Natl Acad Sci USA 2005;102:14197–14202.
15 Guatelli-Steinberg D, Reid DJ, Bishop TA: Did the lateral enamel of Neanderthal anterior teeth grow differently from that of modern humans? J Hum Evol 2007;52:72–84.
16 Hillson S, Bond S: Relationship of enamel hypoplasia to the pattern of tooth crown growth: a discussion. Am J Phys Anthropol 1998;104:89–103.

M. Christopher Dean, PhD
Cell and Developmental Biology, University College London, Anatomy Building
Gower Street
London, WC1E 6BT (UK)
Tel. +44 2076790462, E-Mail ucgacrd@ucl.ac.uk

Koppe T, Meyer G, Alt KW (eds): Comparative Dental Morphology.
Front Oral Biol. Basel, Karger, 2009, vol 13, pp 74–79

Amelogenin Evolution and Tetrapod Enamel Structure

Thomas G.H. Diekwisch[a] · Tianquan Jin[a] · Xinping Wang[a] · Yoshihiro Ito[a] · Marcella Schmidt[a] · Robert Druzinsky[a] · Akira Yamane[b] · Xianghong Luan[a]

[a]Brodie Laboratory for Craniofacial Genetics, University of Illinois at Chicago, Chicago, Ill., USA; [b]Department of Pharmacology, Tsurumi University, Yokohama, Japan

Abstract

Amelogenins are the major proteins involved in tooth enamel formation. In the present study, we have cloned and sequenced four novel amelogenins from three amphibian species in order to analyze similarities and differences between mammalian and non-mammalian amelogenins. The newly sequenced amphibian amelogenin sequences were from a red-eyed tree frog (*Litoria chloris*) and a Mexican axolotl (*Ambystoma mexicanum*). We identified two amelogenin isoforms in the Eastern red-backed salamander (*Plethodon cinereus*). Sequence comparisons confirmed that non-mammalian amelogenins are overall shorter than their mammalian counterparts, contain less proline and less glutamine, and feature shorter polyproline tripeptide repeat stretches than mammalian amelogenins. We propose that unique sequence parameters of mammalian amelogenins might be a pre-requisite for complex mammalian enamel prism architecture.

Amelogenin is the major protein component (90%) of the mammalian enamel protein matrix [1, 2]. A series of genetic, antisense, knockout, and crystal growth studies of the recent decade have established amelogenin's pivotal role in the control of enamel crystal growth and enamel formation [3–6]. While amelogenins are not the only proteins in the developing enamel matrix, they have nevertheless been attributed a major role in the growth of elongated enamel crystals [3–6]. In previous studies from our laboratory, we have characterized the structure of the amelogenin-rich enamel protein matrix [4, 7, 8] and its functional changes related to enamel crystal growth [4]. Our studies have established a functional relevance for the structured amelogenin matrix to control enamel crystal growth [4] and a close relationship between crystal nucleation and changes in matrix configuration during initial crystal formation [8].

In recent years, our knowledge of the enamel protein composition and function of non-mammalian vertebrates has seen significant progress. Amelogenin sequences from reptilian and amphibian teeth have been published [9, 10], and biochemical and immunohistochemical studies have enhanced our knowledge of enamel protein homologies between different vertebrate species [12–16]. Immunohistochemical findings have confirmed earlier reports on a predominance of enamelins in shark enameloid, compared to a predominance of amelogenins in reptilian and mammalian enamel [16]. Based on molecular phylogenetic studies, the amelogenin signal peptide in exon 2 has been linked to a similar region of the SPARC, and suggesting that exon 2 was duplicated to amelogenin approximately 630 mya [11]. According to homology analyses, intermediaries in the molecular evolution of amelogenins from SPARC were SPARCL1, enamelin, and ameloblastin [17].

In the present study, we have generated sequence data for four novel amphibian amelogenins to analyze and compare key parameters of mammalian and non-mammalian amelogenins. In addition, we have documented prismatic mammalian enamel with non-prismatic reptilian and amphibian microarchitecture.

Materials and Methods

Source and Isolation of the Genomic DNA
For the present analysis, three amphibian species were chosen: a red-eyed tree frog *(Litoria chloris)*, a Mexican axolotl *(Ambystoma mexicanum)* and an Eastern red-backed salamander *(Plethodon cinereus)*. Amphibians were euthanized according to approved guidelines by the UIC animal care committee. The genomic DNA was isolated using GenElute™ Mammalian Genomic DNA Miniprep (Sigma-Aldrich Co., St. Louis, Mo., USA) following the manufacturer's instruction. The isolated genomic DNA was kept at –80°C for future use.

RNA Isolation
RNA isolation was performed as previously reported [10]. Briefly, amphibian jaws were removed and immediately frozen in liquid nitrogen and homogenized. The homogenized teeth tissue in TRI AGENT® reagent (Sigma) was mixed with 0.2 ml of chloroform and shaken vigorously for 30 s. The mixture was centrifuged at 12,000 *g* for 20 min at 4°C after 10 min incubation at room temperature. The aqueous phase was mixed with equal volume of isopropanol, and then centrifuged at 12,000 *g* for 20 min at 4°C. The pellet was washed with 70% ethanol and dissolved in DEPC-treated H_2O. The isolated RNA was kept at –80°C for future use.

cDNA Synthesis, Cloning and Sequencing
The sequencing strategy was as described by Wang et al. [10, 18]. Briefly, primers were selected based on sequence analysis of amelogenin genes from selected species with the aid of Lasergene software (DNASTAR Inc., Madison, Wisc., USA). The consensus sequences among different species were chosen as primers for amplification of novel amelogenin genes. Reverse transcriptase reaction was performed using SuperSript™ II Reverse Transcriptase (Invitrogen, Carlsbad, Calif., USA). The PCR products were purified with QIAquick® Gel Extraction Kit (Qiagen Inc., Valencia, Calif., USA), ligated to pGEM®-T Easy vector, and transformed into JM109 competent cells (Promega, Madison, Wisc., USA). Transformants were picked up, and cultivated in LB medium containing ampicillin at a final concentration of 50 µg/ml. Recombinant plasmids were isolated with Wizard® *Plus* SV Minipreps DNA Purification System (Promega), identified by enzyme digestion with *Eco*RI, and sequenced using the ABI 377 sequencer (Northwoods DNA Inc., Becida, Minn., USA). Three colonies were selected and sequenced 2 times from both orientations with either T7 or SP6 primers.

Sequence Analysis
The four newly discovered amelogenin amino acid sequences were deduced from the nucleotide sequence, and sequence analyses were performed to identify its features. Sequences were manually aligned to represent optimum interspecies homology.

Scanning Electron Microscopy
For scanning electron microscopy, frog, squamate, and mammalian teeth from the du Brul collection at the University of Illinois were longitudinally sectioned and etched using EDTA (ethylenediaminetetraacetic acid). Etched enamel surfaces were then analyzed using a Joel JSM-6320F at the UIC RRC laboratory.

Results

Higher Complexity of Tooth Enamel Microstructure in Mammals versus Amphibians/Reptilians

Scanning electron micrographs illustrated prismatic organization of enamel microstructure in all mammals investigated (human, *Homo sapiens*; pig, *Sus scrofa*; steer, *Bos Taurus*; Virginia opossum, *Didelphis virginiana*; and La Plata river dolphin, *Pontoporia blainvillei*), while there was no prismatic enamel structure in most squamates and amphibians (e.g. green iguana, *Iguana iguana*; leopard frog, *Rana pipiens*). The spiny-tailed lizard (*Uromastyx maliensis*) was unique among squamates as its enamel is prismatic (fig. 1).

Four Novel Amphibian Amelogenin Genes Confirm Conserved Elements of Tetrapod Amelogenins

For this study, RNA was isolated from the tooth-bearing elements of three amphibian species, a red-eyed tree frog, *Litoria chloris*; a Mexican axolotl *Ambystoma mexicanum*; and an Eastern red-backed salamander *Plethodon cinereus*. Based on RNA extracts, four novel amphibian amelogenin cDNAs were cloned, sequenced, and deposited in Genbank (red-eyed tree frog, *Litoria chloris*, accession number DQ069788; Mexican axolotl *Ambystoma mexicanum*, accession number DQ069791; and two amelogenin isoforms from the Eastern red-backed salamander *Plethodon cinereus*, accession numbers DQ069789 and DQ069789). Translated amino acid sequences are listed and aligned in figure 2. In this alignment, sequence elements encoded by exons 2, 3 and 5 were highly conserved, while portions of the exon 6 encoded domain were highly variable among species (fig. 2). Nevertheless, also in the exon 6 encoded regions, unique polyproline tripeptide repeats were fairly conserved (fig. 2).

Mammalian amelogenins are longer, contain more proline and glutamine, and have more proline-tripeptide repeats that their amphibian/reptilian counterparts.

In order to determine differences between mammalian and non-mammalian amelogenin proteins, a number of key parameters such as length, amino acid composition, and the number of proline repeats were analyzed based on amelogenin sequences presented in figure 2. Comparing mammalian and non-mammalian amelogenins, the overall amelogenin length increased by 9%, the proline content increased by 21.5%, the glutamine content increased by 31.3%, and the number of proline-tripeptide repeats increased by 72.8% (fig. 3).

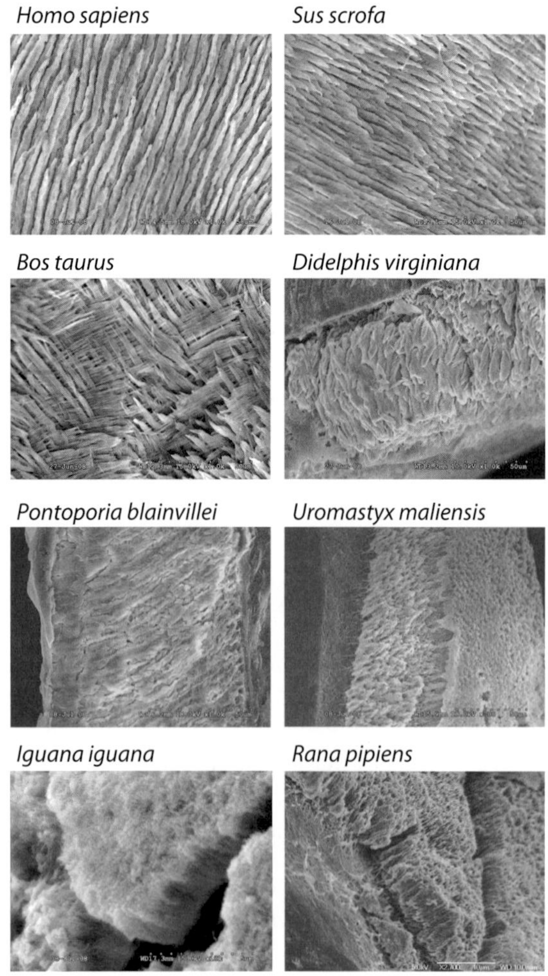

Fig. 1. Enamel prisms and the non-mammalian/mammalian transition. Scanning electron micrographs resolve long and parallel enamel prisms in omnivores (human, *Homo sapiens*, and pig, *Sus scrofa*). Note the pronounced plywood structure in ruminants (steer, *Bos taurus*) and marsupials (Virginia opossum, *Didelphis virginiana*). The enamel layer of dolphins (La Plata river dolphin, *Pontoporia blainvillei*) is fairly thin for eutherians and consists mostly of radial enamel. The spiny-tailed lizard (*Uromastyx maliensis*) is unique among squamates as its enamel is prismatic. In most squamates (e.g. green iguana, *Iguana iguana*) and amphibians (e.g. leopard frog, *Rana pipiens*) the enamel is devoid of prisms.

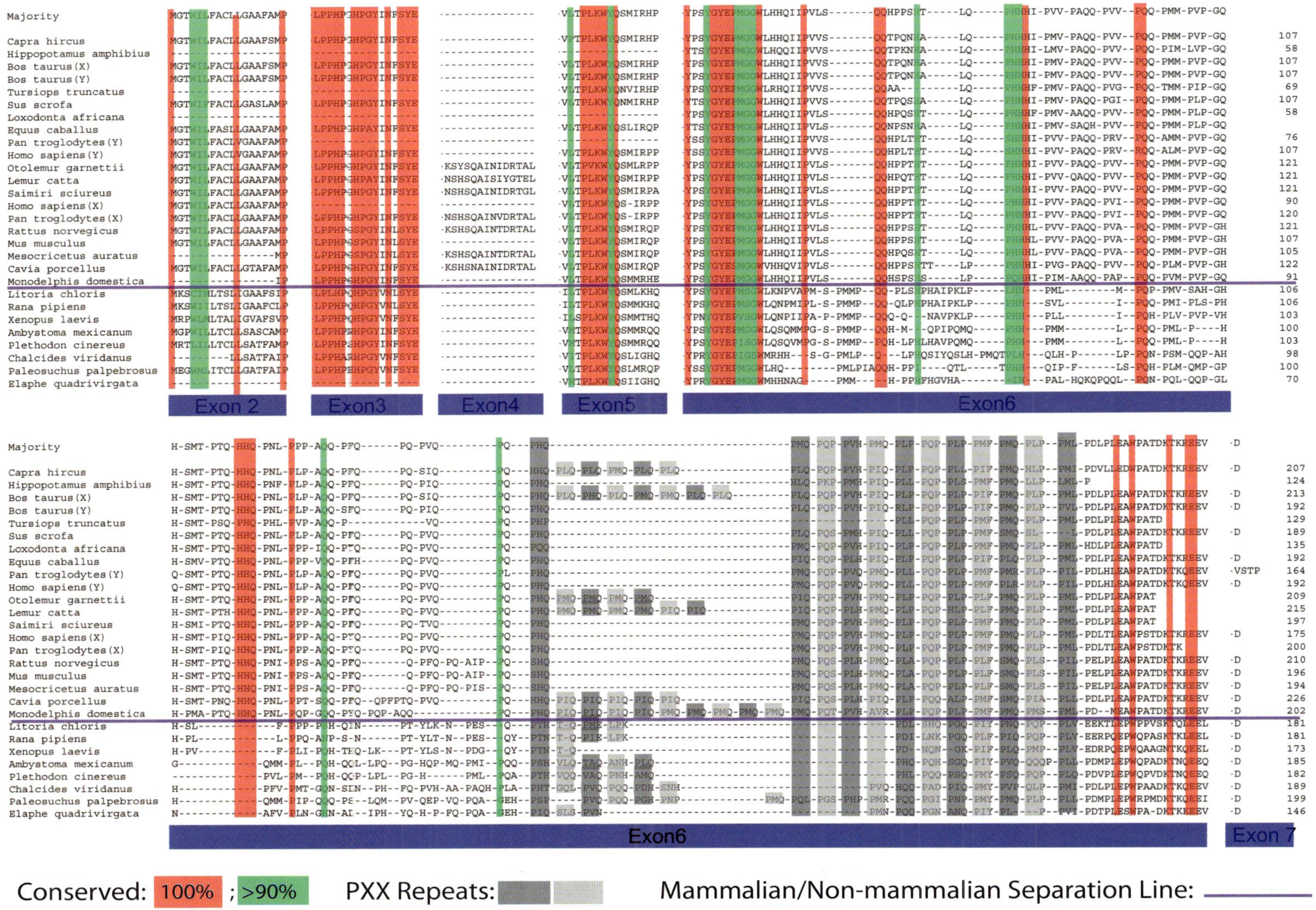

Fig. 2. Conserved regions and evolutionary 'hotspots' among tetrapod amelogenins. Note the highly conserved sequence elements in encoded by exons 2, 3, and 5 (marked in red and green). In contrast, the region encoded by exon 6 reveals significant differences among species. Conserved areas of polyproline tripeptide repeats are labeled in gray. This alignment features four newly discovered amphibian amelogenins (red-eyed tree frog, *Litoria chloris*, accession number DQ069788; Mexican axolotl *Ambystoma mexicanum*, accession number DQ069791; Eastern red-backed salamander *Plethodon cinereus*, accession numbers DQ069789 and DQ069789). The leopard frog *(Rana pipiens)* amelogenin sequence was reported earlier by our group (#AY695795).

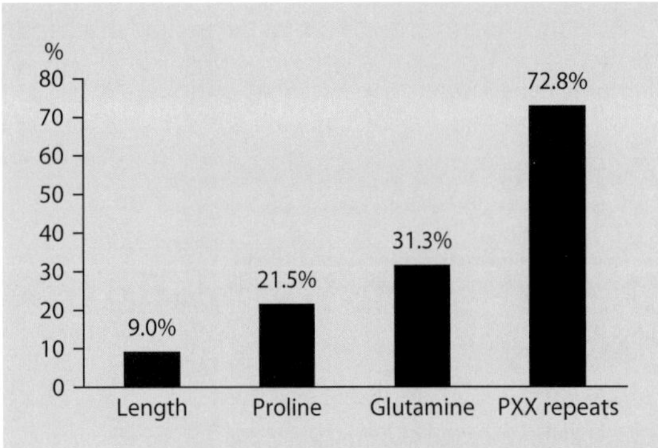

Fig. 3. Differences between mammalian and non-mammalian amelogenin sequence parameters (% increases). A numerical comparison of key parameters of the translated amelogenin protein sequence between mammalian and non-mammalian species revealed that the overall amelogenin length was increased by 9%, the proline content increased by 21.5%, the glutamine content increased by 31.3%, and the number of proline-tripeptide repeats increased by 72.8%.

Discussion

Already studies by Bonass et al. [19] pointed toward a functional significance of the amelogenin center domain as an evolutionary 'hotspot' by identifying seven tandem repeats of a section of nucleotides with the consensus sequence CTGCAGCCC. Further support for the importance of the central amelogenin domain for enamel crystal growth has been provided by transgenic studies documenting that LRAP (a small amelogenin-derived peptide containing most of the A- and B-domains) fails to rescue an amelogenin null phenotype [20]. Recent studies have linked the emergence of genes with new functions to gene duplication and alternative splicing [21, 22] providing theoretical support for a novel evolutionary mechanism by which mammalian amelogenin may have evolved through tandem exon duplication and substitution alternative splicing. According to evolutionary studies, the rapid evolution of the central amelogenin domain is primarily accomplished by insertions of PXX or PXQ tripeptide motifs [19], with both proline and glutamine causing structural rigidity of the newly added tripeptide complexes [23]. These studies suggest that amelogenin evolution is associated with significant alterations in the physicochemical properties of the amelogenin molecule.

Here we have confirmed that amphibian amelogenins are overall shorter than their mammalian counterparts [9], contain less proline and less glutamine, and most significantly, feature shorter polyproline tripeptide repeat stretches than mammalian amelogenins. Moreover, there is ample evidence for a simple, prismless enamel microarchitecture in most amphibians and reptilians, while mammalian enamel is often organized into prisms that frequently form a plywood structure [24]. While direct evidence for a relationship between amelogenin gene structure and enamel prism architecture has not yet been established, we suggest that the increased length and changed composition of mammalian amelogenins provides a basis for an organized protein matrix that might promote increased enamel crystal length and prismatic architecture. Especially the presence of polyproline repeat motifs would provide a potential explanation for the increased rigidity of mammalian amelogenin protein assemblies and thus a mechanism for an orderly growth of long and parallel apatite crystals in complex prism patterns. In addition, the dolphin with its thin and radial enamel and its short polyproline amelogenin repeat motifs might provide one example of a 'link' species, in which the lack of typical mammalian amelogenin characteristics are associated with a reduction in mammalian enamel features.

References

1 Termine JD, Belcourt AB, Christner PJ, Conn KM, Nylen MU: Properties of dissociatively extracted fetal tooth matrix proteins. I. Principal molecular species in developing bovine enamel. J Biol Chem 1980a; 255:9760–9768.

2 Fincham AG, Lau EC, Simmer J, Zeichner-David M: Amelogenin biochemistry – form and function; in Slavkin H, Price P (eds): Chemistry and Biology of Mineralized Tissues. Amsterdam, Excerpta Media, 1992, pp 187–201

3 Lagerstrom M, Dahl N, Nakahori Y, Nakagome Y, Backman B, Landegren U, Pettersson U: A deletion in the amelogenin gene (AMG) causes X-linked amelogenesis imperfecta (AIH1). Genomics 1991;10:971–975.

4 Diekwisch T, David S, Bringas P Jr, Santos V, Slavkin HC: Antisense inhibition of AMEL translation demonstrates supramolecular controls for enamel HAP crystal growth during embryonic mouse molar development. Development 1993;117:471–482.

5 Gibson CW, Yuan Z-A, Hall B, Longenecker G, Chen E, Thyagarajan T, Sreenath T, Wright JT, Decker S, Piddington R, Harrison G, Kulkarni AB: Amelogenin-deficient mice display an amelogenesis imperfecta phenotype. J Biol Chem 2001;276:31871–31875.

6 Iijima M, Moriwaki Y, Wen HB, Fincham AB, Moradian-Oldak J: Elongated growth of octacalcium phosphate crystals in recombinant amelogenin gels under controlled ionic flow. J Dent Res 2002;81:69–73.

7 Diekwisch TGH, Berman BJ, Gentner S, Slavkin HC: Initial enamel crystals are not spatially associated with mineralized dentine. Cell Tissue Res 1995;279:149–167.

8 Diekwisch TG: Subunit compartments of secretory stage enamel matrix. Connect Tissue Res 1998;38:101–111; discussion 139–145.

9 Toyosawa S, O'hUigin C, Figueroa F, Tichy H, Klein J: Identification and characterization of amelogenin genes in monotremes, reptiles, and amphibians. Proc Natl Acad Sci USA 1998; 95:13056–13061.

10 Wang X, Ito Y, Luan X, Yamane A, Diekwisch TG: Amelogenin sequence and enamel biomineralization in *Rana pipiens.* J Exp Zoolog B Mol Dev Evol 2005;304:177–186.

11 Delgado S, Casane D, Bonnaud L, Laurin M, Sire JY, Girondot M: Molecular evidence for precambrian origin of amelogenin, the major protein of vertebrate enamel. Mol Biol Evol 2001;18: 2146–2153.

12 Slavkin HC, Diekwisch T: Evolution in tooth developmental biology: of morphology and molecules. Anat Rec 1996; 245:131–150.

13 Slavkin HC, Diekwisch TGH: Molecular strategies of tooth enamel formation are highly conserved during vertebrate evolution. Ciba Found Symp 1997;205:73–80; discussion 81–84.

14 Kogaya, Y: Immunohistochemical localisation of amelogenin-like proteins and type I collagen and histochemical demonstration of sulphated glycoconjugates in developing enameloid and enamel matrices of the larval urodele (*Triturus pyrrhogaster*) teeth. J Anat 1999;195:455–464.

15 Ishiyama M, Mikami M, Shimokawa H, Oida S: Amelogenin protein in tooth germs of the snake *Elaphe quadrivirgata*, immunohistochemistry, cloning and cDNA sequence. Arch Histol Cytol 1998;61:467–474.

16 Satchell PG, Anderton X, Ryu OH, Luan X, Ortega AJ, Opamen R, Berman BJ, Witherspoon, DE, Gutmann JL, Yamane A, Zeichner-David M, Simmer JP, Shuler CF, Diekwisch TGH: Conservation and variation in enamel protein distribution during tooth development across vertebrates. Mol Dev Evol 2002;294:91–106.

17 Kawasaki K, Weiss KM: Mineralized tissue and vertebrate evolution: the secretory calcium-binding phosphoprotein gene cluster. Proc Natl Acad Sci USA 2003;100:4060–4065.

18 Wang X, Fan JL, Ito Y, Luan X, Diekwisch TG: Identification and characterization of a squamate reptilian amelogenin gene: *Iguana iguana.* J Exp Zoolog B Mol Dev Evol 2006;306:393–406.

19 Bonass WA, Kirkham J, Brookes SJ, Shore RC, Robinson C: Isolation and characterization of an alternatively-spliced rat amelogenin cDNA: LRAP – a highly conserved, functional alternatively-spliced amelogenin? Biochim Biochphys Acta 1994;1219:690–692.

20 Chen E, Yuan ZA, Wright JT, Hong SP, Li Y, Collier PM, Hall B, D'Angelo M, Decker S, Piddington R, Abrams WR, Kulkarni AB, Gibson CW: The small bovine amelogenin LRAP fails to rescue the amelogenin null phenotype. Calcif Tissue Int 2003;73:487–495.

21 Kondrashov FA, Koonin EV: Origin of alternative splicing by tandem exon duplication. Hum Mol Genet 2001;10:2661–2669.

22 Sankoff D: Gene and genome duplication. Curr Opin Genet Dev 2001;11:681–684.

23 Anishetty S, Pennathur G, Anishetty R: Tripeptide analysis of protein structures. BMC Struct Biol 2002;2:9.

24 Sander PM: Presmless enamel in amniotes: terminology, function, and evolution; in Teaford MF, Smith MM, Ferguson MWJ (eds): Development, Function and Evolution of Teeth. Cambridge, Cambridge University Press, 2000, pp 92–106.

Prof. Thomas G.H. Diekwisch
Brodie Laboratory for Craniofacial Genetics, UIC College of Dentistry
801 South Paulina
Chicago, IL 60612 (USA)
Tel. +1 312 413 9683, Fax +1 312 996 6044, E-Mail tomdkw@uic.edu

Koppe T, Meyer G, Alt KW (eds): Comparative Dental Morphology.
Front Oral Biol. Basel, Karger, 2009, vol 13, pp 80–85

Microstructure of Dental Hard Tissues and Bone in the Tuatara Dentary, *Sphenodon punctatus* (Diapsida: Lepidosauria: Rhynchocephalia)

J.A. Kieser[a] · T. Tkatchenko[a] · M.C. Dean[c] · M.E.H. Jones[c] · W. Duncan[a] · N.J. Nelson[b]

[a]Faculty of Dentistry, University of Otago, Dunedin, and [b]School of Biological Sciences, Victoria University, Wellington, New Zealand; [c]Research Department of Cell and Developmental Biology, UCL, University College London, London, UK

Abstract

The Tuatara, *Sphenodon*, is a small reptile currently restricted to islands off the coast of New Zealand where it feeds mainly on arthropods. A widely held misconception is that '*Sphenodon* does not have real teeth' and instead possesses 'serrations on the jaw bone'. One hatchling and one adult dentary were examined under SEM. Two longitudinal ground sections 100-μm thick were prepared through a lower canine tooth and its supporting tissues. There was clear evidence of aprismatic enamel (primless enamel) containing dentine tubules crossing the EDJ, dentine, cementum and a basal-bone attachment. Enamel increments averaged ~3 μm/day and extension rates were ~30 μm/day. The base of the tooth consisted of basal attachment bone that graded from few cell inclusions to lamella or even Haversian-like bone with evidence of remodeling. A string of sclerosed pulp-stone like structures filled the pulp chamber and were continuous with the bone of attachment. Bone beneath the large central nutrient mandibular (Meckel's) canal was quite unlike lamella bone and appeared to be fast growing and to contain wide alternating cell-rich and cell-free zones. Bone cells were rounded (never fusiform) and had few, if any, canaliculi. The dentine close to the EDJ formed at about the same rate as enamel but also contained longer period increments ~100 μm apart. These were spaced appropriately for monthly lunar growth bands, which would explain the basis of the banding pattern observed in the fast growing basal bone beneath the Meckel's canal.

The tuatara *Sphenodon punctatus*, often referred to as a living fossil, is the sole extant member of the order Rhynchocephalia [1, 2], the sister taxon of Squamata (lizards, snakes, amphisbaenians, and mosasaurs). Both groups originated around 240–250 mya, but it was the Rhynchocephalia that were most prevalent during the Early Mesozoic when they attained a global distribution [2] and radiated into a number of distinct groups [3]. In stark contrast, the surviving tuatara is threatened and restricted to wildlife sanctuaries and about 35 islands off New Zealand [4–6].

The dentition of the tuatara is characterised by the presence of two fixed parallel rows of teeth on each side of the upper jaw, i.e. the maxillary and the palatine rows, hence making it the only extant amniote with an enlarged palatine tooth row [2, 7]. The teeth of the lower jaw or dentary occlude within the space between the upper tooth rows and shear forwards after jaw closure [8]. All known fossil taxa possess this additional palatine tooth row, but there is variation in both the shape of the teeth and type of jaw movement employed during food processing [3].

A persistent myth that may stem from remarks made by Colenso [9], is that *Sphenodon* has no teeth, but instead possess serrations on the jaw bone [4]. Teeth in fact are present but fixed to the crest of the jaw bone (acrodonty) so that the boundary between the tooth and jaw bone is almost indistinct [7, 10]. This is compounded by the fact that enamel is quickly worn away such that teeth often comprise only dentine, which further blurs the tooth-jaw boundary. Previous work on tuatara enamel described it as being prismless and columnar [11–14] but there is certainly scope for further studies.

The teeth of tuatara are not replaced, but after hatching, new teeth are added to the posterior end of the tooth-row as it increases in length [10, 16]. One exception to this is the large anterior caniniform teeth that replace some of the hatchling dentition during the juvenile period [10]. This system of tooth growth and implantation seems to have been acquired early on in the group's history and may have been required for their initial radiation [16–20]. However, the exact nature of the tooth tissues that exist as well as the tissues involved in tooth growth and replacement has remained unclear. For example, additional hard tissues deposited around the tooth bases during life have been referred to both as secondary bone [21, 22] and secondary dentine [18]. Detailed histological comparisons with acrodont lizards (e.g. *Uromastyx*, [23]) also remain largely unexplored.

Materials and Methods

Two specimens were examined: a hatchling, approximately 12–15 months of age and an adult female whose age was unknown. Specimens were obtained from Victoria University with the necessary permission as well as Maori consultation.

Scanning Electron Microscopy (SEM)
The external surfaces of premaxillary, maxillary and dentary tooth rows from both specimens were scanned whole, after preparation as described by Boyde [24]. Specimens were also cut longitudinally and coronally to permit imaging of internal surfaces.

Light Microscopy (LM)
Undermineralised ground sections were prepared as follows; a sample from both an adult and a hatchling was dehydrated and embedded in self-polymerising methyl-methacrylate resin at 55°C and 36 psi for 10 min in a pressure container. The embedded specimens were then sectioned using an R330 diamond wheel on a Struers Accutom-50 precision cut-off saw, glued to plastic slides using cyanoacrylate glue, ground and polished using the Struers TegraPol-21/TegraForce 5 system to a final grit size of 4,000 and specimen thickness of 100 μm. Slides were surface-stained with one part MacNeal's tetrachrome (methylene blue, azure II and methyl violet) followed by two parts toluidine blue. Two additional ground sections were prepared from a canine tooth and examined in transmitted polarised light.

Results

The SEM images clearly showed two different structures forming the coronal part of the tooth. A thin outer layer covered the entire tooth structure and appeared to have a different density to the thin layer of tissue covering it. The pulp cavity of the tooth (fig. 1) and the neurovascular canal connecting adjacent teeth (Meckel's canal) were clearly visible. At higher magnifications we were able to identify the bulk of the tooth as dentine, which was covered with an enamel cap. The tooth base was surrounded by a bone of attachment of the type previously described by Smith and Hall [25] and compared by them with cementum. It shows the presence of peri-odontoblastic space with small cross-bridges and branched tubules (fig. 1, inset).

Light-microscopic observations on the teeth allowed us to describe other details about the various hard tissues. The enamel was aprismatic (prismless *sensu* Sander [14]) and had a layered appearance with increments that measured 2.5–3.5 μm apart and these rates matched those in dentine at the EDJ very closely. The enamel contained dentine tubules that changed their orientation at the EDJ but which were restricted to the

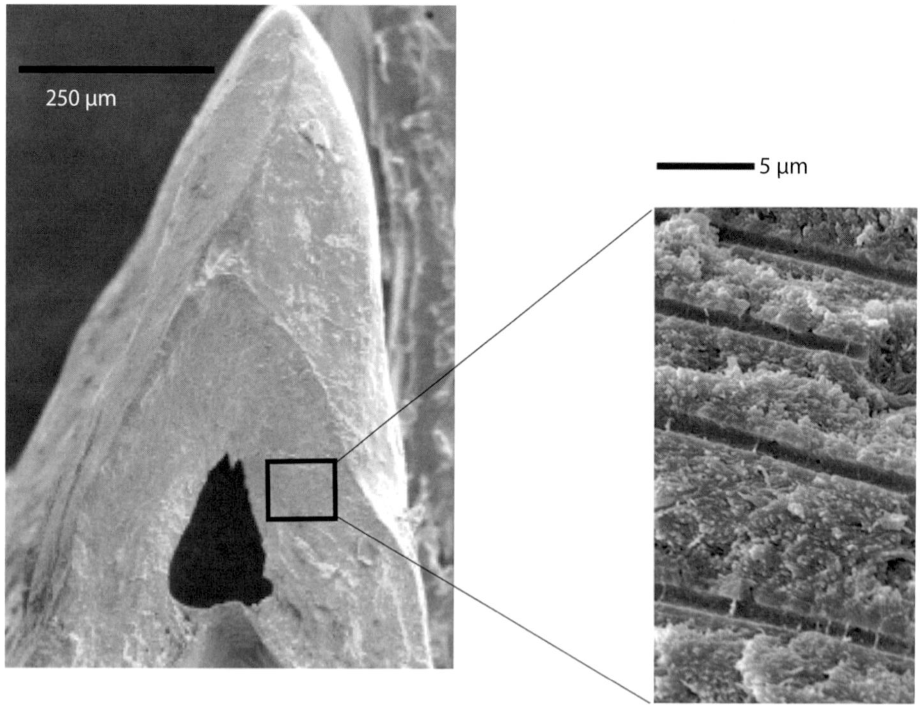

Fig. 1. Scanning electron micrograph of an immature tuatara tooth in coronal section showing a central pulp cavity surrounded by dentine capped thinly by enamel. Inset: Higher power view of branched tubular dentine.

inner enamel. On the assumption that the increments were daily it was possible to calculate the rate at which ameloblasts would have covered the dentine cap during development (the so-called extension rate). The shallow angle of inclination of the layers to the EDJ and the fact that 20–25 increments could be counted along a 700–740 µm length of the EDJ suggests an extension rate of the order of 30 µm/day, which is similar to that observed in, for example, ungulates and may, therefore, represent a typical rate for fast-forming teeth. The dentine was characteristically tubular with 'S-shaped' tubule paths and with many secondary branching tubules in the cervical regions. However, gradually, there was loss of tubules and increased sclerosis towards the basal bone of attachment (fig. 2, 3). Seven accentuated markings existed within a ~700-µm thickness of coronal dentine, which at a daily rate of formation of 3.0 µm/day would suggest a near-monthly rhythm in dentine formation.

Closer towards the bone of attachment there were increasing numbers of cell inclusions. The cementum layer overlying the dentine apical to the CEJ appeared to contain fewer cells than the surrounding bone. Some (maybe older) pulp chambers were filled with a string of 'bead-like' pulp stones. These contained no cells or tubules but consisted only of a swirl of concentric mineralised fibres. They appeared to coalesce with the lamella bone and occasional Haversian-like system at the base of the tooth attachment. All bone (and 'dentine') cell lacunae were always rounded with few if any cell processes and never really fusiform.

The bone of attachment immediately beneath the tooth was lamellar in structure and showed

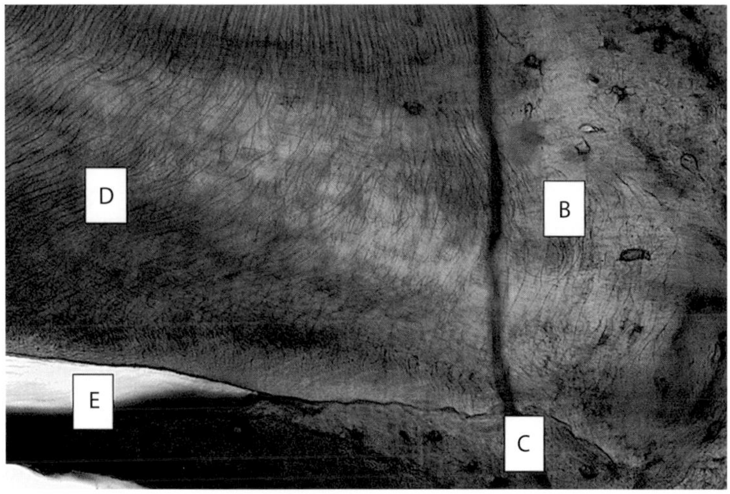

Fig. 2. Light microscope image of an adult tuatara tooth showing cementum C, enamel E, dentine D and bone B.

Fig. 3. Light microscope section of an adult tuatara tooth showing cell-free dentine (right), cells within dentine at the base of the tooth (middle) and basal cellular bone (left). Scale bar = 0.2 mm.

evidence of Howship's lacunae along the margin beneath the tooth. This suggests remodeling of basal attachment bone had taken place, perhaps at the time tooth formation was completed and the tooth became ankylosed into a resorbed and circumscribed region of the crestal bone of the jaw.

The bone of the dentary beneath the neurovascular canal (Meckel's canal) appeared quite different to the lamella-like bone of attachment immediately beneath the tooth. Widely spaced, and presumably fast-growing increments of bone that may again be near-monthly, could be seen sweeping out towards each side of the jaw. There were well-defined and wide alternating zones of cell-rich and cell-free tissue with clear shifts in fibre orientation between these zones in some places. Where they were concentrated, the bone cell lacunae were rounded and none showed any sign of canaliculi and, therefore, of any cell-to-cell contact.

Discussion

This study demonstrates that typical and well-recognised dental tissues exist in the teeth of tuatara and that the nature of tooth implantation on the dentary is via a crestal bone of attachment (acrodonty). Each of the dental tissues, enamel, dentine and cementum, show obvious resemblances to those known from other animals [12–15]. This is important for two reasons. First it clarifies the confusion as to whether the teeth in tuatara were composed of bone or tooth tissue but second, and more importantly, it confirms that the dental tissues were capable of responding dynamically to tooth wear and to function through the usual processes of secondary dentine formation and perhaps pulp-stone formation. The presence of dentine tubules in enamel raises the possibility that enamel, as well as dentine, was capable of reacting to temperature and, when exposed through to the tubules in the inner enamel, to changes in osmotic pressure in the same way dentine does.

The nature of the various kinds of bone in the dentary of tuatara make an interesting comparison with that known from other reptiles and that known in mammals [12–15]. Small rounded cells with few if any cell processes are typical of all bone in the tuatara dentaries examined here. Therefore the hard tissues around the teeth is more correctly referred to as 'secondary bone'. Continued growth of the dentary must take place at a symphysis or growth centre and the evidence from this study is that the nature of the bone at these sites and at the continually growing region of the dentary beneath Meckel's canal is quite different to that immediately beneath the tooth. Fast-growing bone in the basal dentary with zones of cell inclusions appears to be distinct from the lamella bone that shows evidence of remodeling to accommodate the implantation of fully grown teeth.

Interestingly, *Sphenodon* is often thought of as a slow animal because of its metabolism and general movements [4]. This study is the first evidence for hard tissue turnover in tuatara and provides a platform for assessing tooth implantation in the tuatara's fossil relatives, particularly those taxa that are not fully acrodont [2, 17, 19, 20].

Acknowledgment

We thank Moya Smith for discussion and assistance with exact identification of the hard tissues involved.

References

1 Gauthier JA, Estes R, de Queiroz K:1988. A phylogenetic analysis of the Lepidosauromorpha; in Estes R, Pregill G (eds): Phylogenetic Relationships of the Lizard Families: Essays Commemorating Charles L. Camp. Stanford, Stanford University Press, 1988, pp 15–98.
2 Evans SE: At the feet of dinosaurs: the early history and radiation of lizards. Biol Rev 2003;78:513–551.
3 Jones MEH: The evolution of skull shape and feeding strategy in *Sphenodon* and other Rhynchocephalia (Diapsida: Lepidosauria). J Morphol 2008;269:945–966.
4 Mlot C: Return of the tuatara. Sci News 1997;152:300–301.
5 Gaze P: Tuatara recovery plan no. 47, 2001–2011. Te Papa Atawhai, Department of Conservation, 2001.
6 Mitchell NJ, Kearney MR, Nelson NJ, Porter WP: Predicting the fate of a living fossil: how will global warming affect sex determination and hatching phenology in tuatara? Proc R Soc [B] 2008;275:2185–2193.
7 Günther A: Contribution to the anatomy of *Hatteria* (*Rhynchocephalus*, Owen). Phil Trans R Soc 1867;157:1–34.
8 Gorniak GC, Rosenberg HI, Gans C: Mastication in the Tuatara, *Sphenodon punctatus* (Reptilia: Rhynchocephalia): structure and activity of the motor system. J Morphol 1982;171:321–353.
9 Colenso W: Notes on the bones of a species of *Sphenodon* (*S. diversum*, Col) apparently distinct from the species already known. Trans Proc NZ Inst 1886;18:118–123.
10 Robinson PL: How *Sphenodon* and *Uromastix* grow their teeth and use them; in Bellairs Ad'A, Cox CB (eds): Morphology and Biology of the Reptiles. London, Academic Press, 1976, pp 43–64.
11 Carlson SJ, Bartels WS: Ultrastructural complexity in reptilian tooth enamel. Geol Soc Aust Abstr 1986;18:558.
12 Carlson SJ: Vertebrate dental structures; in Carter JD (ed): Skeletal Biomineralization: Pattern, Processes and Evolutionary Trends. New York, Van Nostrand Reinhold, 1990, vol 1, chap 21, pp 531–556.
13 Carlson SJ: Vertebrate dental structures. Part 9; in Carter JD (ed): Skeletal Biomineralization: Pattern, Processes and Evolutionary trends. New York, Van Nostrand Reinhold, 1990, vol 2. pp 73–79, plates 176–192.

14 Sander PM: The microstructure of reptilian tooth enamel: terminology, function, and phylogeny. Münch Geowissensch Abhandl [A] 1999;38:1–102.

15 Sander PM: Prismless enamel in amniotes: terminology, function, and evolution; in Teaford MF, Smith MM, Ferguson MWJ (eds): Development, Function, and Evolution of Teeth. Cambridge, Cambridge University Press, 2000, pp 92–106.

16 Reynoso VH: Growth patterns and ontogenetic variation of the teeth and jaws of the Middle Jurassic sphenodontian *Cynopshenodon huizachalensis* (Rept: Rhynoceph.). Can J Earth Sci 2003;40:609–619.

17 Evans SE: Tooth replacement in the Lower Jurassic lepidosaur *Gephyrosaurus bridensis*. Neues Jahrb Geol Paläontol Abh 1985;1985:411–420.

18 Fraser NC: New Triassic sphenodontids from South West England and a review of their classification. Palaeontology 1986;29:165–186.

19 Whiteside DI: The head skeleton of the Rhaetian sphenodontid *Diphydontosaurus avonis gen. et sp. nov.*, and the modernising of a living fossil. Phil Trans R Soc 1986;B312:379–430.

20 Fraser NC, Shelton CG: Studies of the tooth implantation in fossil tetrapods using high-resolution X-radiography. Geol Mag 1988;125:117–122.

21 Harrison HS: Development and succession of the teeth in *Hatteria punctata*. Q J Microsc Sci 1901;44:161–213.

22 Harrison HS: *Hatteria punctata*, its dentitions and its incubation period. Anat Anz 1901b;.20:145–158.

23 Cooper JS, Poole DFG: The dentition and dental tissues of the agamid lizard *Uromastyx*. J Zool 1973;169:85–100.

24 Boyde A: The development of enameloid structure. Proc R Soc Med 1967;60:923–928.

25 Smith MM, Hall BK: Development and evolutionary origins of vertebrate skeletogenic and odontogenic tissues. Biol Rev 1990;65:277–373.

Prof. J.A. Kieser
Faculty of Dentistry, University of Otago
Dunedin (New Zealand)
Tel. +64 03 479 7083, Fax +64 03 479 7070, E-Mail jules.kieser@stonebow.otago.ac.nz

Koppe T, Meyer G, Alt KW (eds): Comparative Dental Morphology.
Front Oral Biol. Basel, Karger, 2009, vol 13, pp 86–92

Temporal Nature of Periradicular Bands ('Striae Periradicales') on Mammalian Tooth Roots

Tanya M. Smith[a] · Donald J. Reid[b]

[a]Department of Human Evolution, Max Planck Institute for Evolutionary Anthropology, Leipzig, Germany; [b]Department of Oral Biology, Newcastle University, Newcastle upon Tyne, UK

Abstract

Periradicular bands, or fine circumferential lines on tooth roots, have received attention recently due to their prominence on hominin fossils and their potential utility for informing studies of root formation. In 1938, Komai and Miyauti [Dtsch Zahn Mund Kieferheilkd 1938;5:791–795] demonstrated that periradicular bands are related to dentine growth rather than cementum, suggesting that they were equal to accentuated lines in dentine ('dentine lamellae' or 'contour lines'). More recent indirect evidence from band spacing on primate roots suggests that they are temporally equal to other long-period lines in enamel (Retzius lines, perikymata) and dentine (Andresen lines). One of the main complications in understanding the relationship between Andresen lines and periradicular bands is the layer of cementum found on erupted teeth, which often obscures bands. Here we present both direct and indirect evidence that periradicular bands are temporally equivalent to internal long-period lines in the enamel and dentine. A sample of modern human teeth showing periradicular bands and accentuated rings was externally notched, molded, and sectioned; in one instance it was possible to show an equal number of long-period lines (internal Andresen lines and external periradicular bands) between isochrons (internal accentuated lines and external accentuated rings), confirming the temporal equivalence of these features. Furthermore, counts of long-period lines on crown and root surfaces of a Neanderthal anterior dentition showed approximately equal numbers of lines (113 ± 1) between matching hypoplasias and accentuated rings across teeth. Despite their potential for studies of primate root growth, the etiology of these lines in mammalian roots requires further study.

Dental histologists have recognized the existence of incremental lines on tooth crown surfaces and within enamel and dentine for hundreds of years [1]. The structure of enamel is more often studied than dentine, due to anthropological interest in primate enamel thickness, taxonomy, phylogeny, and life history [1, 2]. It is well known that incremental features exist within dental hard tissues that range from sub-daily to annual rhythms [1–4]. Enamel shows long-period markings known as Retzius lines that run from the enamel-dentine junction to the tooth surface, terminating as circumferential ridges on the crown termed perikymata. The direct correspondence between internal and external long-period enamel lines has been demonstrated in the teeth of several primate taxa [5–7]. Dentine also shows long-period lines termed Andresen lines, which are temporally equal to Retzius lines in enamel [1]. In contrast to Retzius lines, it is less clear if Andresen lines terminate on the surface of the dentine, although regularly spaced circumferential features known as periradicular bands are present on mammalian root surfaces (fig. 1).

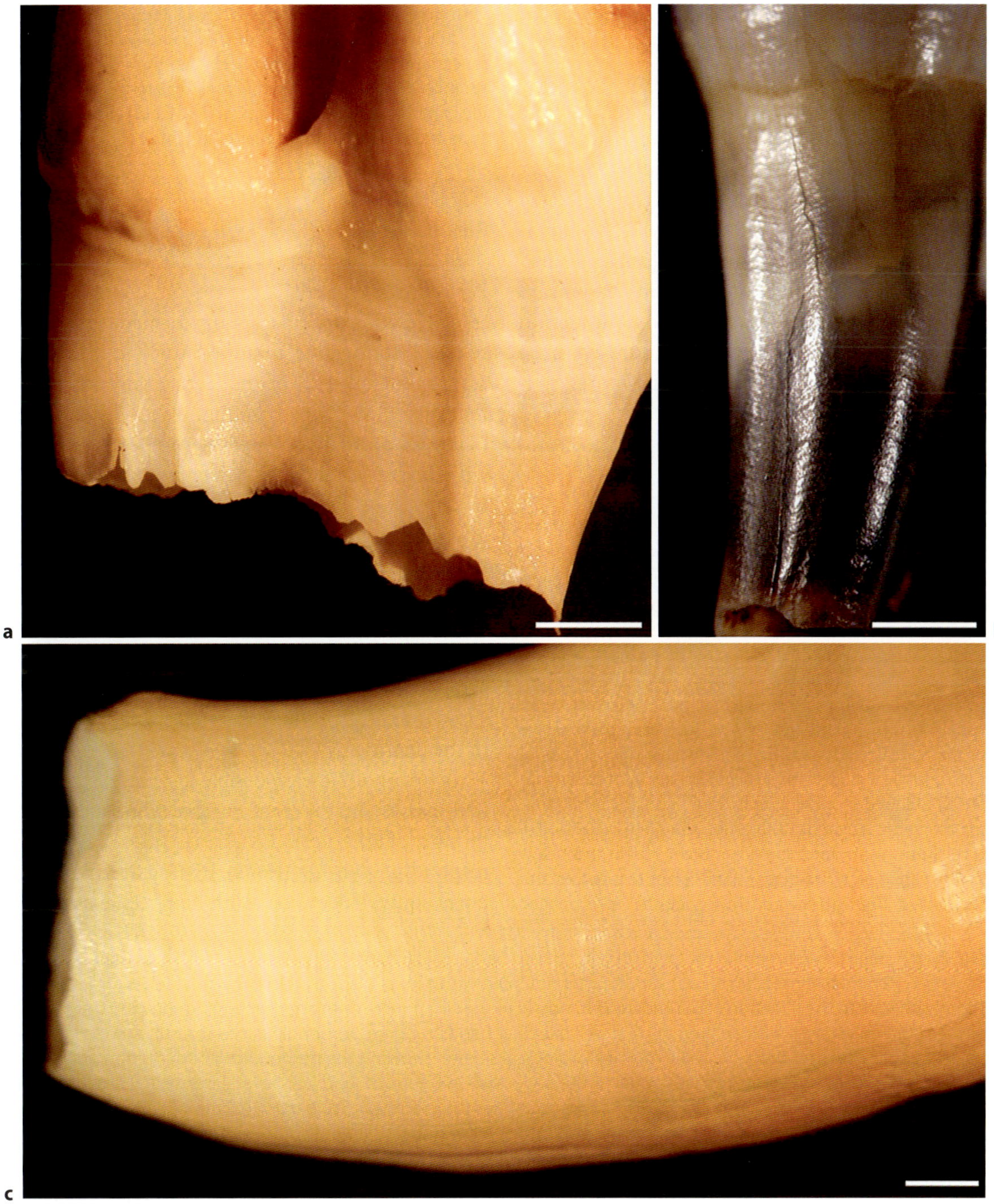

Fig. 1. Fine periradicular bands encircling developing tooth roots in a reindeer molar (**a**), Neanderthal premolar (**b**), and chimpanzee canine (**c**). Particularly marked accentuated rings (parallel to regular periradicular bands) are likely caused by an unknown physiological stressor [12, 13]. Scale bars = 2 mm.

Komai and Miyauti [8] note that Fujita suggested that these external root lines be termed 'striae periradicales', which translates as 'depressions or furrows around the root'. (Here we refer to these features by the contemporary convention 'periradicular bands' [1, 9–11].) Additional irregular rings also encircle tooth roots that represent non-specific stress markers [12], which are similar to hypoplasias on crowns [13], here termed 'accentuated rings'. This study aims to present data on the relationship between long-period lines (Retzius lines, perikymata, Andresen lines) and periradicular bands, permitting more accurate assessment of root formation.

Materials and Methods

Approximately 100 clinically extracted teeth were obtained from European dentists; the sample was primarily comprised of developing and fully formed premolars. Notes were made on the presence of accentuated rings and periradicular bands observed under low magnification stereomicroscopy. Accentuated rings were distinguished from periradicular bands as the former were typically more widely spaced, slightly discolored, and/or slightly raised or depressed relative to the root surface. Impressions of five roots were made with Struers Repliset impression material, followed by photography and histological sectioning along the long axis of the root (detailed in Reid et al. [14]). Roots were notched with a razor above and below distinct areas of periradicular bands prior to molding and sectioning. Impressions and thin sections were imaged with stereomicroscopy and polarized light microscopy, respectively, to compare periradicular bands on the external aspect of teeth with internal Andresen lines in the corresponding region. Root regions were registering by matching external accentuated rings (and notches) with internal accentuated lines (and profiles of notches).

Results

Periradicular bands are rare on clinically extracted teeth, particularly fully formed teeth with residual connective tissue or thick cementum. Approximately 15% of the teeth examined were found to show some banding; only ~5% showed bands for more than 3/4 of the root length. Roots with broad accentuated rings were most likely to show adjacent fine periradicular bands, most often in the middle of the root. Histological sections revealed that, even in those few teeth that expressed clear periradicular bands, Andresen lines were rare in dentine below the cervix, particularly in the lower 2/3 of the root. Roots that showed clear external accentuated rings often showed a corresponding internal accentuated line in the dentine.

In one tooth it was possible to relate periradicular bands on the root surface with an equal number of long-period Andresen lines in the corresponding dentine (fig. 2), demonstrating temporal equivalence. The tooth had been extracted

Fig. 2. a Human lower premolar root with evenly spaced periradicular bands encircling the tooth root (blue arrows). Scale bar = 2 mm. **b** Impression of root surface after notching with a razor blade (green arrows). Notches are oriented slightly oblique to the direction of periradicular bands. Scale bar = 1 mm. **c** Tooth after sectioning showing the paired notches (green arrows), with an area of clear periradicular bands running perpendicular to the cut surface (diagonal through the image). Scale bar = 0.5 mm. **d** Histological section showing the perpendicular orientation of figures 2b and 2c. Notches (green arrows) are equal to the lower pair in figure 2b (and 2c). Several horizontal accentuated lines may be seen, as well as regularly spaced Andresen lines (blue arrows). Scale bar = 0.5 mm. **e** Magnified image of the negative impression shown in figure 2b. The same pair of notches are indicated with green arrows, followed by three horizontal accentuated lines (light blue, red, and yellow). The distances between accentuated lines and the number of periradicular bands in the two intervals are indicated to the right of the brackets. Numbers represent the average of several counts taken on the original tooth and negative impression. Scale bar = 0.5 mm. **f** Magnified polarized image of the histological section in figure 2d, showing the same pair of notches (green arrows), as well as the three corresponding accentuated lines. The distances between accentuated lines and the number of Andresen lines in the two intervals are indicated to the right of the brackets. The number of periradicular bands and Andresen lines between the accentuated lines is equal, confirming the temporal equivalence of these features. Scale bar = 0.5 mm. Image reproduced courtesy of *Proceedings of the National Academy of Sciences USA*.

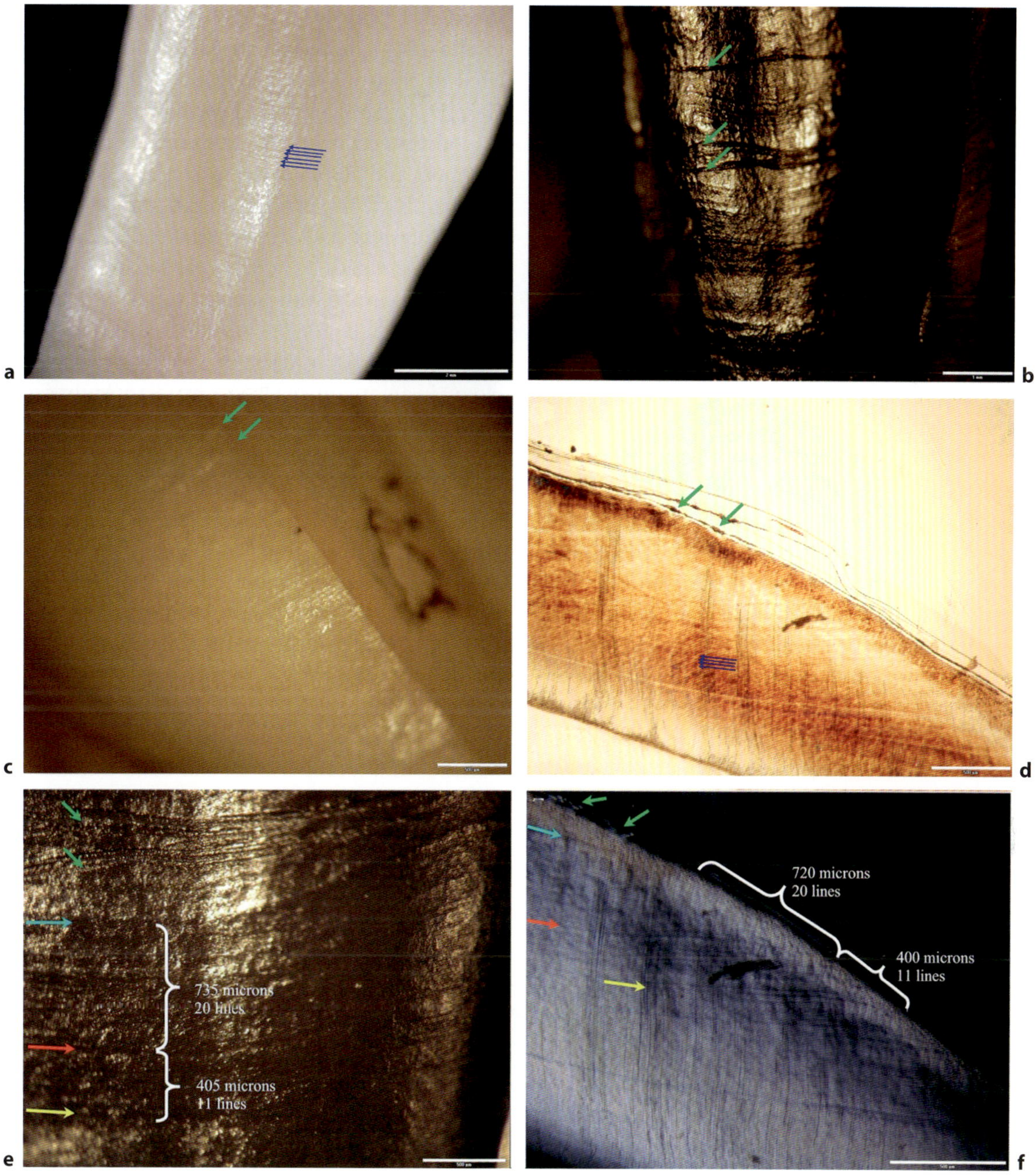
a
b
c
d
735 microns
20 lines
405 microns
11 lines
e
720 microns
20 lines
400 microns
11 lines
f

prior to extensive cellular cementum deposition, which was only apparent histologically in the most apical aspect of the root. The long-period line periodicity in this individual (determined from enamel) was 9 days. The average extension rate in the lower third of the root (fig. 2b–f) was ~3.8 μm/day, and the dentine daily section rate ranged from ~1.5–2.0 μm/day in the first 300 μm deep to the root surface.

Discussion

History of Study

Komai and Miyauti [8] report that Hanazawa and Masaki first described periradicular bands in an experiment published in 1931, regarding them as regular cementum growth lines equivalent to perikymata. Komai and Miyauti [8] illustrate these bands in numerous mammalian taxa, noting variation in band thickness and spacing, and their more frequent appearance near the cervix (relative to root apices). They show that periradicular bands are difficult to see in root areas covered with thick cementum, which they state is due to the fact that periradicular bands are dentine surface configuration lines, commonly overlaid (and obscured) by cementum. These authors conclude that periradicular bands represent the termination of 'contour lines' related to the lamellar structure of dentine. From their illustrations it appears that they are referring to irregular accentuated lines sometimes called contour lines of Owen (see Dean [1]). We advocate that two classes of root surface lines be distinguished: long-period incremental lines commonly termed periradicular bands, and irregular accentuated rings (misleadingly termed root hypoplasias [12]).

Observations of regular concentric lines on mammalian root surfaces have also been made on seal canines [15] and beaver incisors [11]. In the former case they were interpreted to represent annual dentine growth lines formed during the winter, while in the later case they were used to calculate dentine growth rates, which were related to environmental conditions. Rinaldi and Cole [11] review Rinaldi's experimental work that demonstrated circadian periradicular bands in marmots and laboratory rats, stating 'periradicular bands are the external manifestation of von Ebner lines' [p. 290]. However, one difficulty in identifying dentine incremental features is the parallel nature of Andresen and von Ebner's lines (see images in [2, 3]), and it is unclear if these increments are distinguishable (or of separate etiology) in an organism with a circadian Andresen line periodicity.

Newman and Poole [9, 10] were some of the first to suggest that periradicular bands are equivalent to perikymata in primate enamel. Dean [1] provided indirect evidence by documenting the spacing and number of periradicular bands on a *Homo habilis* juvenile's roots (OH 16). He observed that periradicular band spacing did not vary greatly from cervix to apex, but anterior teeth showed more widely spaced bands than posterior teeth. Using a range of likely periodicities and total periradicular band numbers, Dean estimated formation times and extension rates that would result if periradicular bands represented long-period lines, finding a pattern akin to modern great apes. In a similar approach, Berkovitz et al. [16] measured the spacing of periradicular bands in a modern human child, which were similar to extension rates derived from radiographic studies.

Most recently, Smith et al. [17–19] observed and counted these fine lines on several Middle Paleolithic fossil hominin teeth, making predictions of root formation duration and age at death by assuming periradicular bands were equivalent to other long-period lines. In their study of the Scladina Neanderthal's dentition, Smith et al. [17] found equal numbers of perikymata and periradicular bands (113 ± 1 line) between a pair of hypoplasias/accentuated rings on tooth crowns/roots in 5 of 6 anterior teeth (see Smith et al. [17]:

fig. 1, p. 20222). This strongly suggests that these features represent equivalent periods of time, given that varying proportions of crowns and roots yielded essentially equal counts.

Appearance and Etiology
Dean [20] suggested that periradicular bands are difficult to see due to overlying cementum, as well as their more tightly packed and less prominent appearance relative to perikymata. However, a number of unresolved issues remain concerning cement formation [21]. During cementogenesis and prior to tooth eruption, acellular cementum is deposited on a thin layer of unmineralized predentine [22]. It is likely that the clarity of periradicular bands in developing teeth is due to the fact that a significant amount of cellular cementum has yet to be deposited. Given the appearance of periradicular bands on high-resolution developing root impressions, it appears that the dentine surface is not initially blanketed and flattened by cementum. A thin layer of initial acellular cementum may adhere to the dentine surface in a manner that preserves the underlying dentine topography. Periradicular bands are less evident in fully erupted and root complete teeth, which tend to possess more cementum, although they may be seen where the cementum has been removed [8].

In addition to the difficulties associated with resolving periradicular bands, section obliquity may influence the appearance of Andresen lines in histological sections of tooth roots (also noted by Dean [1]), which may explain the direct correlation of Andresen lines and periradicular bands in only a single tooth in this study. Dean [20] also suggested that Andresen lines may be difficult to resolve below the dentine surface due to dentine microanatomy and subsurface mantle dentine formation. However, there is some debate over whether mantle dentine is found along the cementum-dentine junction, or if it is only found near the enamel-dentine junction [22]. In roots the outermost layer of dentine, the hyaline layer, mineralizes after the bulk of radicular dentine. It is unknown how this layer relates to the underlying dentine, although dentine tubules can be identified that are continuous with the deeper dentine [23]. We concur with Dean's [20] statement that the subsurface dentine revealed in histological sections does not typically display incremental features, although we note that it is possible in some instances (e.g. fig. 2f).

In closing, we have demonstrated that periradicular bands on tooth root surfaces are equivalent to long-period Andresen lines in root dentine. Given Dean's [1] demonstration that internal long-period lines in enamel and dentine are equivalent, we suggest that the periodicity of long-period lines on the external surfaces of primate tooth crowns and roots, and within the enamel and dentine, is also equivalent. The same long-period rhythm has also recently been identified in bone [24]. Incremental features on tooth roots may compliment information derived from tooth crowns; however, additional research is needed to understand the diversity of mammalian incremental features, the formation of the interface between dentine and cementum, and the etiology of long-period lines.

Acknowledgements

Chris Dean, Kornelius Kupczik, and Paul Tafforeau are thanked for discussions of this topic, and Jean-Jacques Hublin is acknowledged for his support. Frank Loddemann provided invaluable assistance with German translation, and Paul Tafforeau kindly assisted with synchrotron experiments. Supported by the Max Planck Society and Harvard University.

References

1 Dean MC: The nature and periodicity of incremental lines in primate dentine and their relationship to periradicular bands in OH 16 (*Homo habilis*); in Moggi-Cecchi J (ed): Aspects of Dental Biology: Paleontology, Anthropology and Evolution. Florence, International Institute for the Study of Man, 1995, pp 239–265.

2 Smith TM: Incremental dental development: methods and applications in hominoid evolutionary studies. J Hum Evol 2008;54:205–224.

3 Bromage TG: Enamel incremental periodicity in the pig-tailed macaque: a polychrome fluorescent labeling study of dental hard tissues. Am J Phys Anthropol 1991;86:205–214.

4 Smith TM: Experimental determination of the periodicity of incremental features in enamel. J Anat 2006;208:99–114.

5 Risnes S: Circumferential continuity of perikymata in human dental enamel investigated by scanning electron microscopy. Scand J Dent Res 1985;93:185–191.

6 Kelley J, Smith TM: Age at first molar emergence in early Miocene *Afropithecus turkanensis* and life-history evolution in the Hominoidea. J Hum Evol 2003;44:307–329.

7 Tafforeau PT, Smith TM: Nondestructive imaging of hominoid dental microstructure using phase contrast X-ray synchrotron microtomography. J Hum Evol 2008;54:272–278.

8 Komai S, Miyauti T: Über die Parallelsteifen der Zahnwurzel ('Striae periradicales'). Deutsche Zahn-, Mund- und Kieferheilkunde 1938;5: 791–795.

9 Newman HN, Poole DFG: Observations with scanning and transmission electron microscopy on the structure of human surface enamel. Arch Oral Biol 1974;19:1135–1143.

10 Newman HN, Poole DFG: Dental enamel growth. J Roy Med Soc 1993; 86:61.

11 Rinaldi C, Cole TM: Environmental seasonality and incremental growth rates of beaver (*Castor canadensis*) incisors: implications for palaeobiology. Paleo Paleo Paleo 2004;206:289–301.

12 Teegan WR: Hypoplasia of the tooth root: a new unspecific stress marker in human and animal paleopathology. Am J Phys Anthro 2004;Suppl 38:193.

13 Hillson S, Bond S: Relationship of enamel hypoplasia to the pattern of tooth crown growth: a discussion. Am J Phys Anthro 1997;104:89–103.

14 Reid DJ, Schwartz GT, Dean C, Chandrasekera MS: A histological reconstruction of dental development in the common chimpanzee, *Pan troglodytes*. J Hum Evol 1998;35:427–448.

15 Scheffer VB: Growth layers on the teeth of pinnipedia as an indication of age. Science 1950;112:309–311.

16 Berkovitz BKB, Grigson C, Dean MC: Caroline Crachami, the Sicilian dwarf (1815–1824): was she really nine years old at death? Am J Med Gen 1998;76:343–348.

17 Smith TM, Toussaint M, Reid DJ, Olejniczak AJ, Hublin J-J: Rapid dental development in a Middle Paleolithic Belgian Neanderthal. Proc Natl Acad Sci USA 2007;104:20220–20225.

18 Smith TM, Tafforeau PT, Reid DJ, Grün R, Eggins S, Boutakiout M, Hublin J-J: Earliest evidence of modern human life history in North African early *Homo sapiens*. Proc Natl Acad Sci USA 2007;104:6128–6133.

19 Smith TM, Reid DJ, Olejniczak AJ, Bailey S, Glantz M, Viola B, Hublin J-J: Dental development and age at death of a Middle Paleolithic juvenile hominin from Obi-Rakhmat Grotto, Uzbekistan; in Condemi S, Schrenk F, Weniger G (eds): Neanderthals, Their Ancestors and Contemporaries. Dordrecht, Springer, in press.

20 Dean C: Hominoid tooth growth: using incremental lines in dentine as markers of growth in modern human and fossil primate teeth; in Hoppa RD, FitzGerald CM (eds): Human Growth in the Past. Cambridge, Cambridge University Press, 1999, pp 111–127.

21 Nanci A, Somerman MJ: Periodontium; in Nanci A (ed): Ten Cate's Oral Histology: Development, Structure, and Function. St. Louis, Mosby, 2003, pp 240–274.

22 Schroeder HE: Oral Structural Biology. Stuttgart, Thieme, 1991.

23 Owens PDA: The root surface in human teeth: a microradiographic study. J Anat 1976;122:389–401.

24 Bromage TG, Lacruz RS, Hogg R, Goldman HM, McFarlin SC, Warshaw J, Dirks W, Perez-Ochoa A, Smolyar I, Enlow DH, Boyde A: Lamellar bone is an incremental tissue reconciling enamel rhythms, body size, and organismal life history. Calcif Tissue Int 2009;84:388–404.

Tanya M. Smith, PhD
Department of Human Evolutionary Biology, Harvard University
11 Divinity Avenue
Cambridge, MA 02138 (USA)
Tel. +1 617 496 8259, Fax +1 617 496 8041, E-Mail tsmith@fas.harvard.edu

Koppe T, Meyer G, Alt KW (eds): Comparative Dental Morphology.
Front Oral Biol. Basel, Karger, 2009, vol 13, pp 93–99

Consequences of X-Linked Hypohidrotic Ectodermal Dysplasia for the Human Jaw Bone

H. Lesot[a,b,d] · F. Clauss[a–c] · M.C. Manière[b,c] · M. Schmittbuhl[a–c]

[a]INSERM UMR 977, Faculty of Medicine, and [b]Dental Faculty, University of Strasbourg, and [c]Reference Center for Dental Manifestations of Rare Diseases, University Hospital, Strasbourg, France; [d]International Collaborating Centre in Oro-facial Genetics and Development, University of Liverpool, School of Dental Sciences, Liverpool, UK

Abstract

Mutations of the *Eda* gene, which encodes for ectodysplasin-A1, result in X-linked hypohydrotic ectodermal dysplasia (XLHED). This pathology may lead to severe oligodontia, subsequently requiring implant therapy. Since Eda is suspected to participate in bone development, the jaw bone status was investigated in XLHED patients in order to adjust the surgical protocol. Using computed tomography, densitometric profiles and 3D reconstructions, the bone structure was analyzed and compared to that of control individuals; our results showed that the morphological changes comprised mandibular bone flattening. Craniofacial CT scans showed medullary bone hyperdensity, including in the mandibular symphysis area, where implants must be placed. These alterations in bone structure were also observed in locations where the presence/absence of teeth cannot interfere. If the changes in jaw bone morphology can be a consequence of oligodontia, the changes in bone structure seem to be tooth-independent and suggest a direct effect of the mutation on bone formation and/or remodeling.

The *Eda* gene encodes for ectodysplasin-A1, a type II transmembrane protein with a short intracellular domain, a transmembrane domain, a collagen motif, and a tumor necrosis factor (TNF) ligand motif [1]. Mutations in this gene result in X-linked hypohydrotic ectodermal dysplasia (XLHED, OMIM 305100). Among several other phenotypic alterations, XLHED patients often show severe oligodontia, which then requires dental implant therapy [2, 3]. Since oligodontia in these patients is systematically more severe in the lower jaw, the mandibular bone status was investigated to adjust the surgical protocol, as bone quality and quantity are determining factors in the success of implant therapy [4]. Previous investigations have also shown that Eda is expressed by secreting osteoblasts during embryonic skeletal development [5, 6] and may thus interfere with bone formation. In this study, computed tomography, densitometric profiles and 3D reconstructions were used to investigate jaw bone morphology and structure and to compare them to those in controls. The main results showed mandibular bone flattening and medullary bone hyperdensity, mainly in the mandibular symphysis area, where implants must be placed. However, alterations in bone structure were also observed in locations where teeth cannot interfere, suggesting a direct effect of the mutation on bone formation and/or remodeling. Conversely, the morphological changes observed in the jaw bone appear to be a consequence of oligodontia.

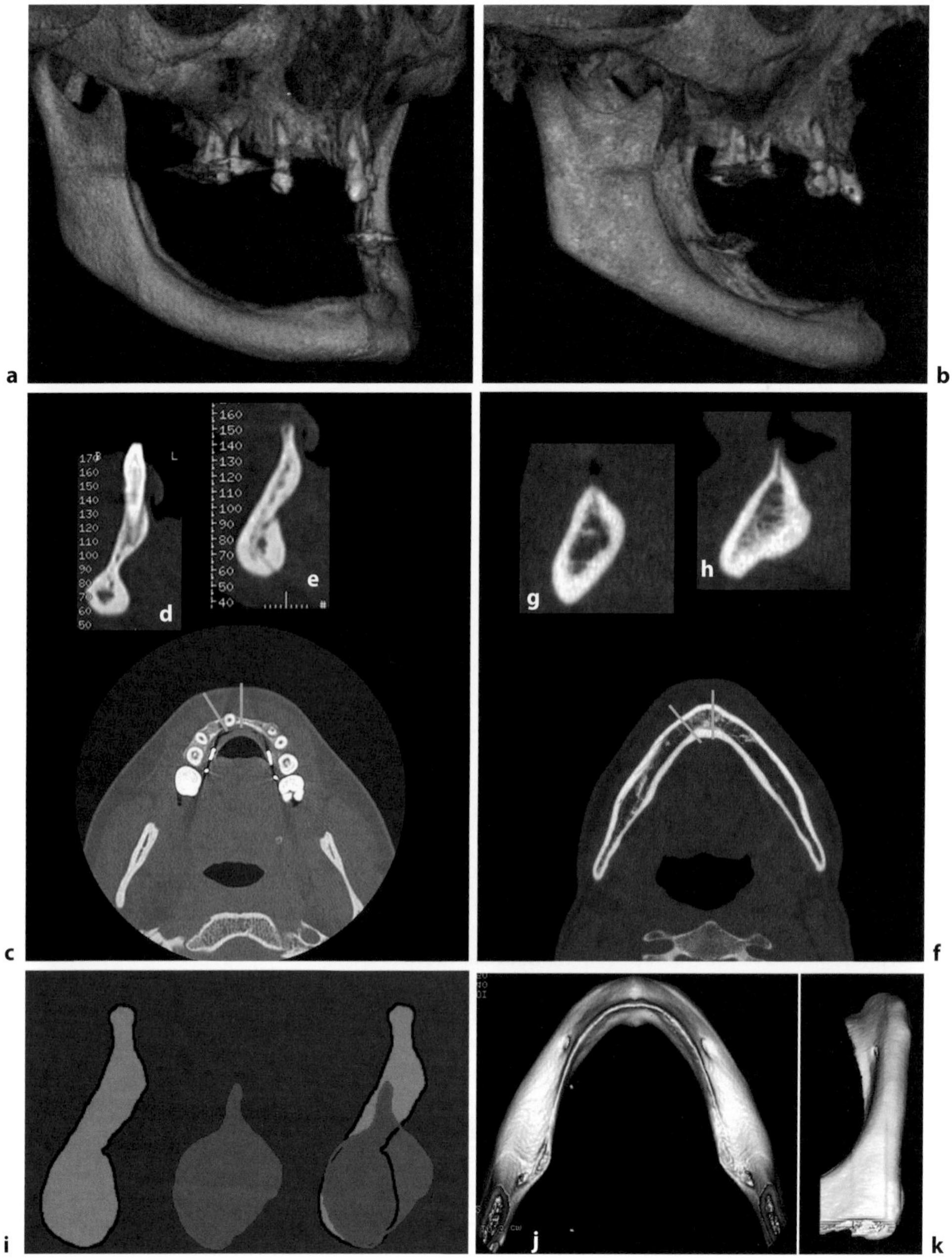

Fig. 1. Mandibular bone morphology in XLHED patients. Three-dimensional reconstructions of an XLHED patient's craniofacial complex (**a, b**). Axial (**c**) and sagittal CT mandibular sections (**d, e**) in a control patient. Axial (**f**) and sagittal (**g, h**) sections in an XLHED patient. The lines in **c** and **f** indicate the positions of the sections shown in **d, e, g** and **h**, respectively. The superposition (right in **i**) of median sagittal sections in the symphysis of XLHED (center in **i**) and control (left in **i**, which corresponds to **e**) patients (**i**) illustrates the change in jaw bone morphology. Upper (**j**) and lateral (**k**) views of an XLHED patient's mandibular 3D reconstruction show the presence of a crest on the upper part of the mandible.

Material and Methods

Recruitment of Patients

Male XLHED patients genotyped for an *Eda* mutation (n = 26) were recruited in the Reference Center for Dental Manifestations of Rare Diseases (University Hospital, Strasbourg) and through multi-disciplinary consultation in the French Reference Center for Genodermatosis (Necker Children's Hospital, Paris). The age of the patients used for this study ranged from 8 to 33 years.

Bone Phenotype Analysis

Bone phenotype analysis was performed using computed-tomography images of the maxilla and the mandible. Bone density was evaluated on craniofacial CT sections by measurements of Hounsfield Units (HU) and based on the Lekholm and Zarb bone density classification [7]. D1 density corresponded to a high bone density (above 850 HU) and D4 to a low bone density (below 500 HU) [4]. Computer-assisted bone densitometric profiles allowed the comparison of the density of XLHED and control bones. These analyzes were performed using Simplant™ (Columbia Scientific, Inc., Columbia, S.C., USA) interactive software specifically designed for the assessment of implant surgery. Bone hyperdensity was visualized on 3D reconstructions, generated from the original CT scan images, using Amira software (version 2.3, TGS, Chelmsford, Mass., USA). Since XLHED is a rare disease, statistical tests could not be performed. However, the parameters investigated in this study in male XLHED patients were compared to those of control male individuals of the same or similar age.

Results

Morphological Changes in the Jaw Bone

In addition to oligodontia, XLHED patients showed a morphologically altered mandibular jaw bone characterized by a flattened shape (fig. 1a, b, g, h). A crest sometimes remained on the upper part of the mandible (fig. 1b, h, i, j; fig. 3d), regardless of the age of the patient (8 years old: fig. 1j, or 33 years old: fig. 1h). Sagittal sections in the CT scans were made in the anterior part of the mouth in XLHED patients (fig. 1g, h) and compared to those of controls (fig. 1d, e). The decrease in jaw bone height was obvious in XLHED patients (fig. 1g, i), and the crest that remained might correspond to the upper part of the jaw bone in controls (compare fig. 1h and fig. 1e; superposition in fig. 1i). Further bone alterations included maxillary hypoplasia and an increased length of the mandibular body (fig. 1b and cephalometric data not shown), suggesting an altered pattern of skeletal growth.

Structural Bone Alterations

Compared to the same anatomical area in control patients, the symphyseal mandibular bone of XLHED patients showed structural defects, including an increased density of the medullary bone (compare fig. 2i, j and d, e) and hypercorticalization (fig. 1h). Computer-aided 3D reconstructions in XLHED patients showed the existence of hyperdense structures in the medullary bone in the symphyseal area (fig. 3e), compared to controls (fig. 3c). These structures were also observed more anteriorly, and were more abundant in XLHED patients (fig. 3f) than in controls (fig. 3b). Furthermore, local variations were observed when comparing the bone density of XLHED patients in the symphyseal area (fig. 2f, i, j) or in the posterior mandibular area (fig. 2f, g, h). Such variations were not observed in control patients (fig. 2a–e). The medullary bone density from the symphyseal area exhibited local variations as observed in axial CT sections (fig. 3a), which ranged from 742 to 1,406 HU.

Axial sections in the CT scans showed an increased density of the cortex of the basal bone in the mandible of XLHED patients (fig. 3g), reaching as high as 1,812 HU, while control values ranged between 1,300 and 1,560 HU. This hypercorticalization was also documented in XLHED patients from CT sagittal slices in the mandibular posterior area (fig. 2g), as compared to the results seen in control CT slices (fig. 2b). In XLHED patients, the densitometric profiles exhibited larger peaks (fig. 2h) than those in controls (fig. 2c), thus suggesting a relative thickening of the cortex.

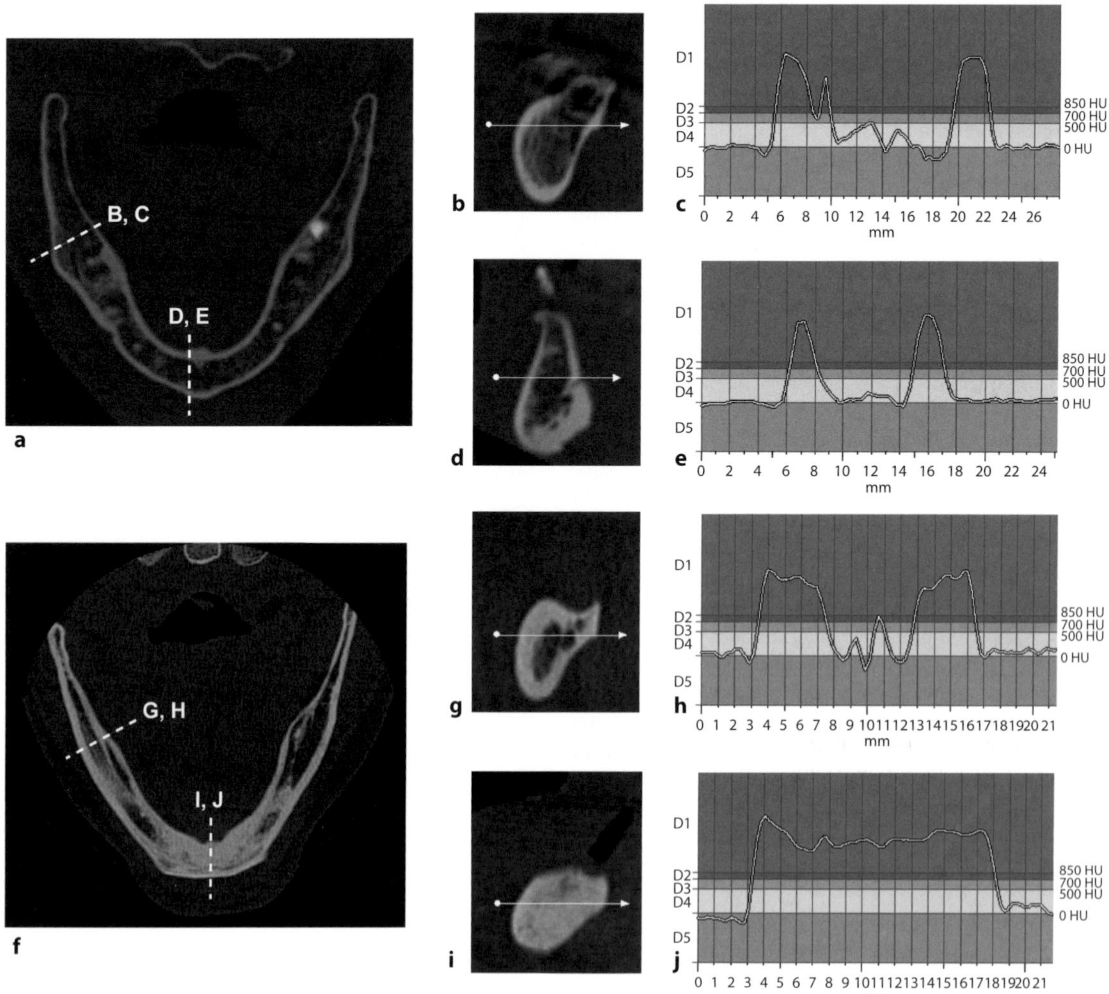

Fig. 2. Mandibular bone density in control (**a**–**e**) and XLHED (**f**–**j**) patients. Axial CT sections of a control subject (**a**) and an XLHED patient (**f**). Cross-sectional slices in the mandibular symphyseal area in the control (**d**) and the XLHED patient (**I**) and the corresponding densitometric profiles (**e**, **j**). Cross-sectional slices in the mandibular posterior area in a control subject (**b**) and XLHED (**g**) patient with the corresponding densitometric profiles (**c**, **h**).

Relationship between Structural Bone Modifications and Tooth Abnormalities

The decrease in the height of the mandibular body is a consequence of tooth agenesis (compare fig. 1g, h in an XLHED patient with fig. 1d, e in a control), while structural defects might be independent of the dental phenotype. To investigate this possibility, measurements were performed in the upper jaw, where teeth were present (fig. 3h, i). There, the density of the cortex increased not only in the alveolar bone, where teeth were present (fig. 3i), but also in the basal bone (fig. 3h), underlying but distant from any teeth (fig. 1d). For the alveolar bone, the density ranged between 991

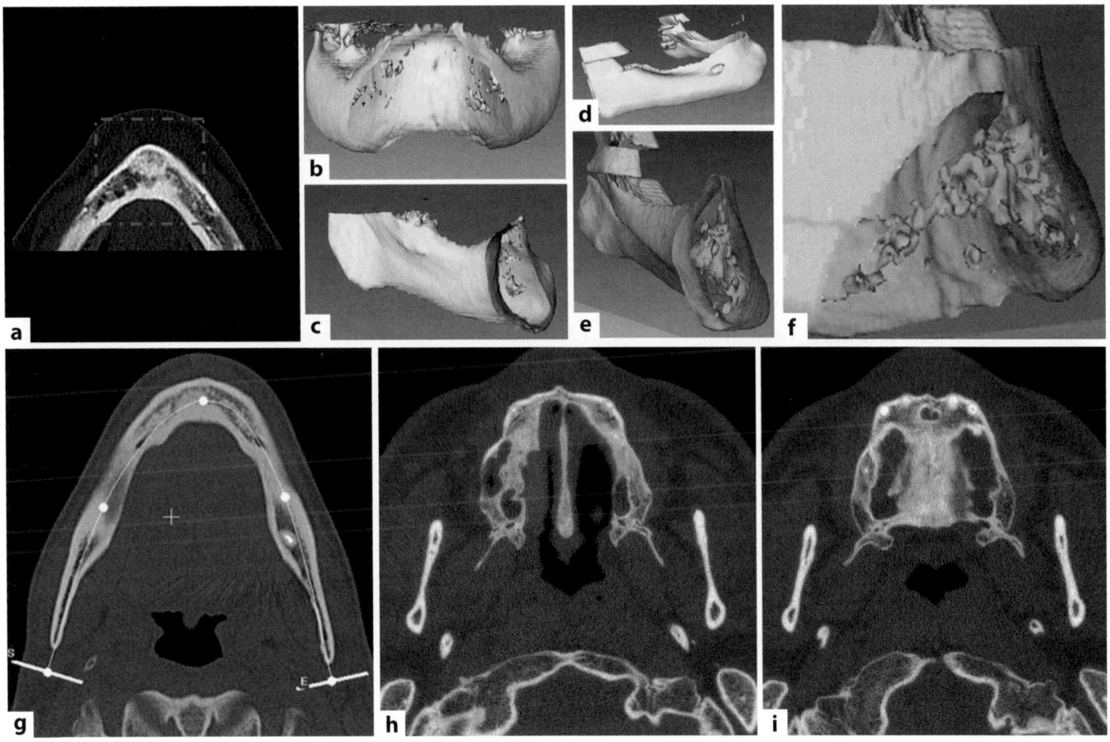

Fig. 3. Bone structure (**a–f**) and density (**g–i**) in control and XLHED patients. Symphyseal mandibular area of an XLHED patient (**a**). Three-dimensional reconstructions of control (**b**, **c**) and XLHED patients (**d–f**) show the presence of hyperdense structures in the mandibular bone. Bone density analysis of the mandibular cortical bone in an XLHED patient (**g**) revealed a value of 1,812 HU compared to a range of 1,300–1,560 HU in controls. Bone density analysis of the maxillary basal bone (**h**) yielded a value of 1,739 HU compared to 835–1,113 HU in controls. Bone density analysis of the maxillary alveolar bone (**i**) resulted in a value of 1,507 HU compared to 810–940 HU in controls.

and 1542 HU in XLHED patients, compared to values ranging between 810 and 940 HU in controls [7]. The values for basal bone density ranged from 1,011 to 1,739 HU in XLHED patients and between 835 and 1,113 HU in controls [7].

Discussion

Mutations in the Eda-A1 gene lead to X-linked hypohydrotic ectodermal dysplasia in humans, which, among other abnormalities, has major consequences for tooth development [8]. XLHED patients show variations in the degree of tooth agenesis and in the location of missing teeth [9]. These variations in the dental phenotype may be caused by epigenetic factors [10, 11], but also by differences in the penetrance of the mutation [12]. Implant therapy is required in about 80% of XLHED patients. Since bone quality and quantity are prognosis factors for the success of implant therapy, the bone phenotype was investigated in a group of XLHED patients.

The XLHED patients included in this study showed both morphological and structural modifications. Mandibular bone hypoplasia was observed in association with tooth agenesis. Similar observations have been made in the Tabby mutant mouse model [13]. Analyzing Tabby mice with a unilateral missing incisor in the lower jaw, it was suggested that the change in alveolar morphology was directly related to the absence of the tooth. However, XLHED patients also exhibited craniofacial dysmorphies, independently of dental abnormalities [3]. Indeed, they showed mandibular pronathism, frontal prominence and growth modification of the cranial base [14, for review see 3 and refs therein]. For this reason, the bone status of these patients was investigated further, revealing changes in bone structure including areas of local increases in medullary bone density and hypercorticalization. Mutations in the Eda-A1 gene have much more dramatic effects on the lower dentition. Attempts were thus made to compare the effects on the bone structure of each jaw individually, in order to determine whether structure alterations might be a direct consequence of the mutation. Human cortical and trabecular bones are heterogeneously mineralized due to local variations in the kinetics of both mineralization and remodeling [15]. Our results showed that in both jaws, there was an increase in basal bone mineralization, independent of the presence or absence of teeth. Bone modifications independent of tooth status were confirmed by comparing the density of the alveolar bone, in contact with teeth, with that of the basal bone, which is distant from teeth. In both cases, the bone density was higher in XLHED patients than in controls, documenting again a direct effect of the mutation on bone structure. Indeed, in humans, Eda is expressed by osteoblasts, suggesting that EDA-A1 may have direct effects on bone formation [5]. Moreover, chondrocytes and hypertrophic chondrocytes do not express Eda, suggesting that the corresponding protein may have more specific functions in membrane ossification than in enchondral ossification processes [5].

Besides the possible role of osteoblasts, osteoclast activity might be altered in XLHED patients and lead to an increase in jaw bone density. The level of parathyroid hormone (PTH), involved in osteoclast differentiation, was found to be decreased in XLHED patients [16]. Further, EDA/EDAR activates the NF-κB intracellular pathway, which is implicated in osteoclastic differentiation and bone remodeling control [17, 18]. Adapter molecules such as TAB2, TAK1 and TRAF6 participate in the transduction of the signal between the complex EDA-ectodysplasin receptor (EDAR) and NF-κB [19]. TRAF6 is involved in ectodermal morphogenesis and later in osteoclastic differentiation and bone remodeling through EDA. Indeed, an osteopetrotic phenotype has been observed in the knock-out mouse model TRAF6–/–, with increased bone radioopacity and altered teeth eruption, characteristic of osteopetrosis [20]. Molecular alterations of the NF-κB pathway are associated with metabolic and structural bone defects as in syndromic ectodermal dysplasia with osteopetrosis linked to NF-κB essential modulator (*NEMO*) mutation [21]. The skeletal phenotype associated with this mutation confirms the involvement of the EDA-NF-κB pathway in bone metabolism. To continue with this study and to evaluate the respective roles of osteoblasts and osteoclasts, in vitro cultures will have to be studied using the Tabby mouse model. However, the present observations illustrate the necessity of adjusting the surgical implantation protocol in light of the morphological change and increased density of the jaw bone in XLHED patients.

Acknowledgement

The authors wish to thank Dr. James Dutt for critical reading of the manuscript.

References

1 Bodmer JL, Schneider P, Tschopp J: The molecular architecture of the TNF superfamily. Trends Biochem Sci 2002;27:19–26.

2 Guckes AD, Scurria MS, King TS, McCarthy GR, Brahim JS: Prospective clinical trial of dental implants in persons with ectodermal dysplasia. J Prosthet Dent 2002;88:21–25.

3 Clauss F, Manière M-C, Obry F, Waltmann E, Hadj-Rabia S, Bodemer C, Alembik Y, Lesot H, Schmittbuhl M: Dento-craniofacial phenotypes and underlying molecular mechanisms in hypohidrotic ectodermal dysplasia (HED): a review. J Dent Res 2008;87:1089–1099.

4 De Oliveira RC, Leles CR, Normanha LM, Lindh C, Ribeiro-Rotta RF: Assessments of trabecular bone density at implant sites on CT images. Oral Surg Oral Med Oral Pathol Oral Radiol Endod 2008;105:231–238.

5 Montonen O, Ezer S, Saarialho-Kere UK, Herva R, Karjalainen-Lindsberg ML, Kaitila I, Schlessinger D, Srivastava AK, Thesleff I, Kere J: The gene defective in anhidrotic ectodermal dysplasia is expressed in the developing epithelium, neuroectoderm, thymus, and bone. J Histochem Cytochem 1998;46:281–289.

6 Tucker AS, Headon DJ, Schneider P, Ferguson BM, Overbeek P, Tschopp J, Sharpe PT: Edar/Eda interactions regulate enamel knot formation in tooth morphogenesis. Development 2000;127:4691–4700.

7 Park HS, Lee YJ, Jeong SH, Kwon TG: Density of the alveolar and basal bones of the maxilla and the mandible. Am J Orthod Dentofacial Orthop 2008;133:30–37.

8 Matalova E, Fleischmannova J, Sharpe PT, Tucker AS: Tooth agenesis: from molecular genetics to molecular dentistry. J Dent Res 2008;87:617–623.

9 Lexner MO, Bardow A, Hertz JM, Nielsen LA, Kreiborg S: Anomalies of tooth formation in hypohidrotic ectodermal dysplasia. Int J Paediatr Dent 2007;17:10–18.

10 Swinnen S, Bailleul-Forestier I, Arte S, Nieminen P, Devriendt K, Carels C: Investigating the etiology of multiple tooth agenesis in three sisters with severe oligodontia. Orthod Craniofac Res 2008;11:24–31.

11 Townsend G, Harris EF, Lesot H, Clauss F, Brook A: Morphogenetic fields within the human dentition: a new, clinically relevant synthesis of an old concept. Arch Oral doi:10.1016/j.archoralbio.2008.06.011.

12 Mostowska A, Kobielak A, Trzeciak WH: Molecular basis of non-syndromic tooth agenesis: mutations of MSX1 and PAX9 reflect their role in patterning human dentition. Eur J Oral Sci 2003;111:365–370.

13 Lesot H, Kristenova P, Haze A, Deutsch D, Peterka M., Peterkova R: Lower incisor in the Tabby mutant mouse; in Landis WL, Sodek J (eds): Proc 8th Conf Chemistry and Biology of Mineralized Tissues, 2004, pp 196–199.

14 Lexner MO, Bardow A, Bjorn-Jorgensen J, Hertz JM, Almer L, Kreiborg S: Anthropometric and cephalometric measurements in X-linked hypohidrotic ectodermal dysplasia. Orthod Craniofac Res 2007;10:203–215.

15 Roschger P, Paschalis EP, Fratzl P, Klaushofer K: Bone mineralization density distribution in health and disease. Bone 2008;42:456–466.

16 Söderholm AL, Kaitila I: Expression of X-linked hypohidrotic ectodermal dysplasia in six males and in their mothers. Clin Genet 1985;28:136–144.

17 Kumar A, Eby MT, Sinha S, Jasmin A, Chaudhary PM: The ectodermal dysplasia receptor activates the nuclear factor-kappaB, JNK, and cell death pathways and binds to ectodysplasin A. J Biol Chem 2001;27:2668–2677.

18 Smahi A, Courtois G, Rabia SH, Döffinger R, Bodemer C, Munnich A, Casanova JL, Israël A: The NF-kappaB signalling pathway in human diseases: from incontinentia pigmenti to ectodermal dysplasias and immune-deficiency syndromes. Hum Mol Genet 2002;11:2371–2375.

19 Morlon A, Munnich A, Smahi A: TAB2, TRAF6 and TAK1 are involved in NF-kappaB activation induced by the TNF-receptor, Edar and its adaptator Edaradd. Hum Mol Genet 2005;14:3751–3757.

20 Lomaga MA, Yeh WC, Sarosi I, et al: TRAF6 deficiency results in osteopetrosis and defective interleukin-1, CD40, and LPS signaling. Genes Dev 1999;13:1015–1024.

21 Döffinger R, Smahi A, Bessia C, et al: X-linked anhidrotic ectodermal dysplasia with immunodeficiency is caused by impaired NF-kappaB signaling. Nat Genet 2001;27:277–285.

Dr. Hervé Lesot
INSERM UMR 977, Faculté de Médecine
11, rue Humann
FR–67085 Strasbourg Cedex (France)
Tel. +33 3 90 24 31 11, Fax +33 3 90 24 35 64, E-Mail Herve.Lesot@odonto-ulp.u-strasbg.fr

Koppe T, Meyer G, Alt KW (eds): Comparative Dental Morphology.
Front Oral Biol. Basel, Karger, 2009, vol 13, pp 100–101

Dental Growth and Development: An Introduction

Inger Kjær

Department of Orthodontics School of Dentistry, Faculty of Health Sciences, University of Copenhagen, Copenhagen, Denmark

Sections on dental ontogeny are among the traditional topics of former International Symposia of Dental Morphology. These sections predominantly dealt with aspects of dental tissue differentiation during tooth development. Even though information about cell and tissue differentiation had a great impact on our knowledge of tooth formation, especially also in fossils, it has to be considered that teeth develop within the craniofacial skeleton. Therefore, the current section on dental growth and development was intended to promote research on dental formation within the craniofacial complex. One keynote lecture and two invited talks reflected this broad area of research, followed by a number of podium and poster presentations.

Tanya M. Smith (Max Planck Institute for Evolutionary Anthropology, Leipzig, Germany) focused in her keynote lecture on new approaches in the study of incremental dental development and how this research can help us to interpret the dental size reduction in *Homo* during the Pleistocene. Karin Binner Bector (University Hospital Copenhagen, Denmark) discussed etiological factors behind human tooth eruption based on both clinical and histological information. Finally, Ivo Lambrichts (Hasselt University at Diepenbeek, Belgium), lectured on immunohistochemical characteristics of neuro-odontoblast and neuro-epithelial interactions. Among others Lambrichts stressed the fact that the sensory innervation might play a regulatory role in maintaining the epithelial rests of Malassez.

Due to strict page limitation, the current section contains a selection of five papers that were presented in this section on dental growth and development. Nevertheless, they resemble excellent approaches in research on dental growth and development.

In the first paper Zhang's group explore the effects of the vitamin D receptor (VDR) pathway on both the enamel and dentin mineralization in a VDR-deficient mouse model. VDR deficiency resulted in different distribution patterns of enamel and dentin mineral deposition. Vitamin D appears to regulate dentin mineralization indirectly, while having a direct influence on enamel mineralization. In a continuously erupting mouse incisor, they observed a reduction in dentin mineralization and early enamel hypermineralization.

Studies of dental asymmetry can provide insights into the biological basis of lateralisation in humans. The results can also assist clinicians to discriminate between normal and abnormal developmental patterns. Based on a large sample of Australian twin pairs Mihailidis and co-workers found evidence for fluctuating asymmetry in primary tooth emergence, especially of lateral incisors. No evidence was found, however, for a genetic contribution to fluctuating asymmetry in tooth emergence. Even though the results also indicated a tendency toward left-side advancement for some teeth, there was no statistical evidence for directional asymmetry.

Growth increments in histological sections are useful markers of tooth formation. While enamel formation rates of permanent teeth vary from 2.5 μm per day at the enamel dentine junction (EDJ) to 6.5 μm per day, little is known about the growth rates of deciduous teeth. In the third paper of this section, Birch and Dean report about the enamel formation rates of human mandibular deciduous teeth of all tooth types. While the enamel of deciduous teeth seems to form at a more constant rate than in permanent teeth, gradients in secretion rates between the EDJ and dental surface and cuspal and cervical regions still exist. Most notably is the observation of Birch and Dean that true enamel hypoplasia, either associated with or without a line, is an equally good measure of stress during deciduous crown formation as well as in permanent crown formation.

Although primates with diets that require higher occlusal forces have tooth roots with significantly larger relative surface areas, tooth root length shows considerably intraspecific variation. Since a number of earlier studies demonstrated, among others, a relationship between facial morphology and root length, Cobb and Baverstock investigate the covariation between tooth root length and the whole craniofacial skeleton using a sophisticated geometric morphometrics approach in *Pan troglodytes*. The observation of correlation between root length variation and facial height rather than length might be due to variation in the eruptive distance associated with compensatory jaw rotation during development.

In the fifth paper, Kelly and co-workers explore factors underlying the differences in age at M1 emergence in great apes by studying tooth root length variation in mandibles with erupting first or second molars. These data complement existing data on crown and root formation times as very important life history markers. Kelly and co-workers observed the longest tooth roots in *Gorilla*, followed by *Pongo* and *Pan*. They suggest that the delay in humans relative to the great apes is due at least in part to a relatively longer period of root formation, resulting from both a lower average root extension rate as well as greater root formation at emergence in humans compared to great apes.

Prof. Inger Kjær
Department of Orthodontics School of Dentistry
Faculty of Health Sciences
University of Copenhagen
Nørre Allé 20
DK--2200 Copenhagen N (Denmark)
Tel. +45 35 32 66 74; E-Mail ik@odont.ku.dk

Koppe T, Meyer G, Alt KW (eds): Comparative Dental Morphology.
Front Oral Biol. Basel, Karger, 2009, vol 13, pp 102–109

Regulation of Enamel and Dentin Mineralization by Vitamin D Receptor

Xueming Zhang[a] · Preston Beck[a] · Firoz Rahemtulla[a] · Huw F. Thomas[b]

Departments of [a]Prosthodontics and [b]Pediatric Dentistry, School of Dentistry, University of Alabama at Birmingham, Birmingham, Ala., USA

Abstract

Background: Vitamin D plays an important role in bone mineralization. Enamel and dentin are two mineralized tissues of different origins that are part of the tooth structure, but the mechanism by which vitamin D regulates the mineralization of these tissues remains unclear. We examined the mineral deposition pattern of enamel and dentin in continuously erupting incisors in a vitamin D receptor (VDR) deficient mouse model to determine the effect of vitamin D receptor pathway on enamel and dentin mineralization. **Methods***:* VDR wild-type mice (VDR+/+) and VDR-deficient (VDR–/–) littermates were sacrificed at 70.5 days of age, and their mandibles were dissected. Immunostaining of biglycan and decorin was used to evaluate the dentin maturation. Micro-computerized tomography (micro-CT) was used to compare the mineral density (MD) of enamel and dentin of the two groups at different regions along the axis of the mandibular incisors. Scanning electronic microscopy (SEM) was employed to examine the ultrastructure of enamel and dentin at the levels corresponding to those examined in the micro-CT studies. Furthermore, an accelerated eruption procedure was performed to exclude the effect of delayed eruption on enamel and dentin mineralization. **Results:** Different mineral deposition patterns of enamel and dentin were observed at different levels of the incisors in the VDR+/+ and VDR–/– groups. Early enamel maturation and mineralization, and dentin hypomineralization were observed in the VDR–/– group. **Conclusion:** Vitamin D affects enamel and dentin mineralization through different mechanisms. It may affect the mineralization of dentin systemically while enamel mineralization may be regulated locally.

The vitamin D pathway is known to systemically manage mineral homeostasis through actions on intestinal absorption and renal re-absorption of calcium and phosphorus, as well as through actions on trapping and mobilizing calcium from mineralized tissues [1, 2]. Timely access to adequate quantities of vitamin D is a vital component of bone mineralization and is necessary for longitudinal maintenance of mineralized tissue in humans and animals. Existing evidence suggests that the stimulation of intestinal absorption of calcium and the inhibition of PTH-induced bone resorption are major factors in the vitamin D pathway affecting osteogenesis [2]. In addition, vitamin D is also reported to function locally by binding to the vitamin D receptors (VDRs) and influencing proliferation of osteoblasts and osteoclasts [3, 4].

Numerous studies utilizing both animal models and human trials have shown that disruption of vitamin D pathway leads to inadequate levels of calcium and phosphorous in circulating plasma. This results in decreased mineralization of skeletal bones, and also has a negative impact on tooth mineralization [5–8]. There is little evidence, however, to show that the same mechanisms are responsible for both altered mineralization of skeletal bone and disrupted mineralization of enamel and dentin. We hypothesized that vitamin D affects bone, enamel, and dentin mineralization through different mechanisms. Thus, if the vitamin D pathway is disturbed, different patterns of mineralization will be observed in these tissues.

Distribution Patterns of Mineral Deposition in Bone, Enamel and Dentin

In order to determine different regulation of vitamin D pathways on bone, dentin and enamel mineralization, we examined the mineral deposition patterns in the continually erupting lower incisors of a VDR knockout mouse model. In this model, a VDR fragment spanning exons 3–5, which encodes the second zinc finger of DNA-binding domain was deleted [9]. Human vitamin D-resistant rickets-like phenotypes were observed in this model, including $1,25(OH)_2D_3$ resistance, hypocalcemia and secondary hyperparathyroidism. Using micro-CT, we observed reduced bone mineral density and osteoporosis-like bone phenotype (fig. 1a). The dentin showed lower mineral density. Multiple radiolucent spots of different sizes were scattered in dentin of the deficient mouse. Large pulp chambers and thinner pulp walls were also observed (fig. 1a). The calcein apposition lines in VDR–/– dentin were diffuse, thus dentin apposition rate could not be determined (fig. 1b). Dentin maturation was delayed and disorganized; dentin and predentin layers were much wider (fig. 1c). In these layers immunostaining showed diffuse expression of biglycan and decorin, which is degraded in mature dentin (fig. 1d). These observations strongly indicate that dentin mineralization and maturation is compromised by VDR deficiency in a manner similar to skeletal bone.

Theoretically, if they shared the same mechanism under the same physical and chemical circumstances, the same patterns of incisal enamel and dentin mineralization should develop [10]. Our next objective was to determine whether enamel follows the same hypomineralization pattern as dentin in VDR-deficient mice. As mouse incisors erupt continuously, multiple mineralization stages of enamel and dentin mineralization are present in a single incisor. We used micro-CT scanning at specified planes along the axis of both VDR+/+ and VDR–/– mouse mandibular incisors to measure the mineral deposition of enamel and dentin at each plane. In the VDR–/– mouse we observed hypermineralization of enamel in the apical region of the incisors (fig. 2a), conversely, at the same observation plane, the dentin showed much less mineral density. SEM observation confirmed early enamel maturation in the corresponding region (fig. 2a). Different distribution patterns of mineral deposition in enamel and dentin were observed between the VDR+/+ and VDR–/– groups (fig. 2b). The distribution pattern of enamel and dentin mineral deposition in VDR–/– mouse model differed significantly from the VDR+/+ group, indicating that VDR deficiency affects enamel and dentin mineralization differently.

Since vitamin D deficiency delays the eruption rate of the incisors [10–12], we believed that delayed eruption may lead to early enamel hypermineralization, and with delayed eruption, the existing enamel would migrate posteriorly. Thus, the level marking the full maturation of the enamel moves under the gingival margin of the incisor and later approaches the apical region of the tooth. To exclude the possible influence that delayed eruption has on enamel mineralization, a published procedure [11, 12] to accelerate the eruption of the lower incisors was performed in both groups. The incisor in one side of each mouse

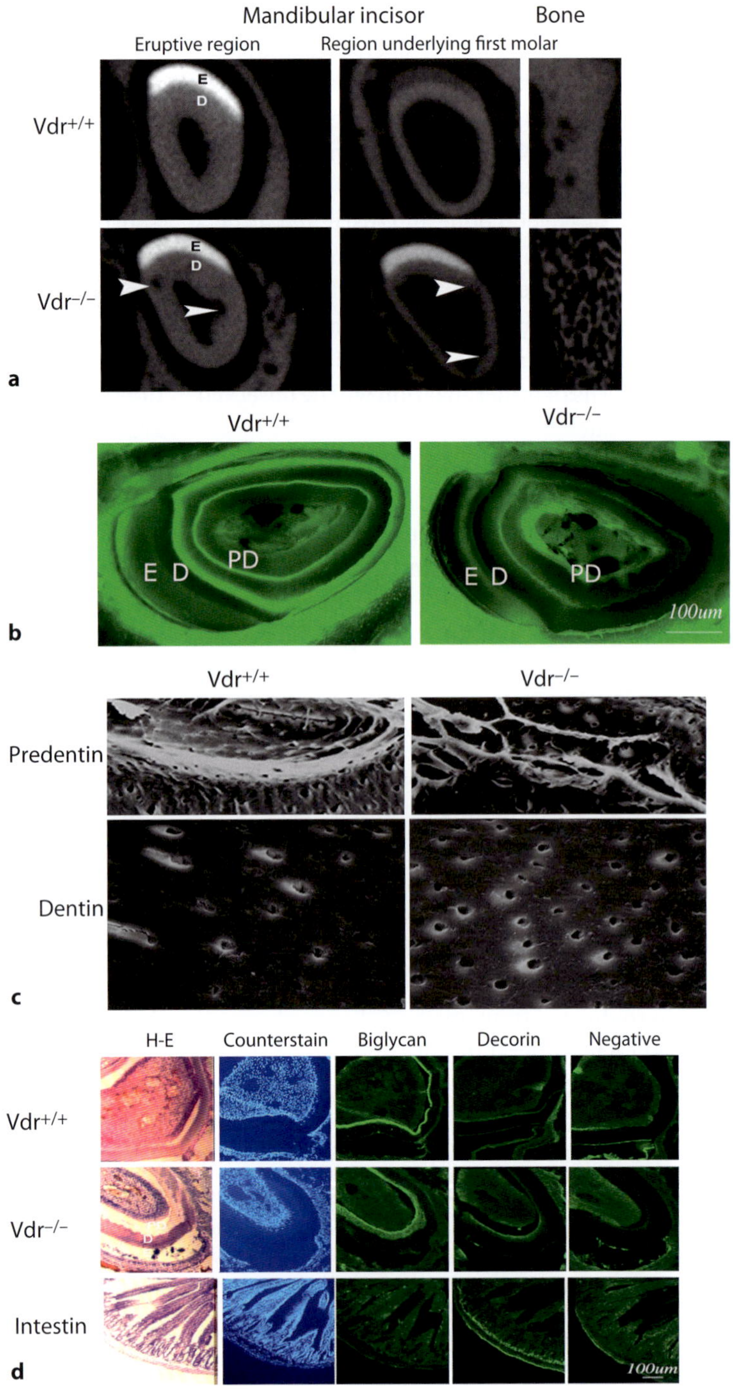

Fig. 1. Dentin hypomineralization observed in 70.5-day-old VDR knockout mice (*vdr–/–*). **a** Micro-CT revealed translucent spots scattered in the dentin of *vdr–/–* mice (arrowheads). The pulp chamber was larger, and the dentin wall was thinner. The high porosity was also observed in the tibia bone of *vdr–/–* mice. **b** Determination of dentin apposition rate in the *vdr+/+* and *vdr–/–* mice. Calcein was double injected at 1-week intervals, and then observed in the following week. At the first molar level of the incisor, double deposition lines were distinct in the *vdr+/+* mice, but they were diffuse in the *vdr–/–* mice. **c** SEM showed that highly disorganized predentin layer in the *vdr–/–* mice. Reduced dentin tubules were also observed. **d** Immunostaining revealed diffuse expression of biglycan and decorin in the erupted region of the VDR deficient mouse incisors. E = Enamel; D = dentin; PD = predentin.

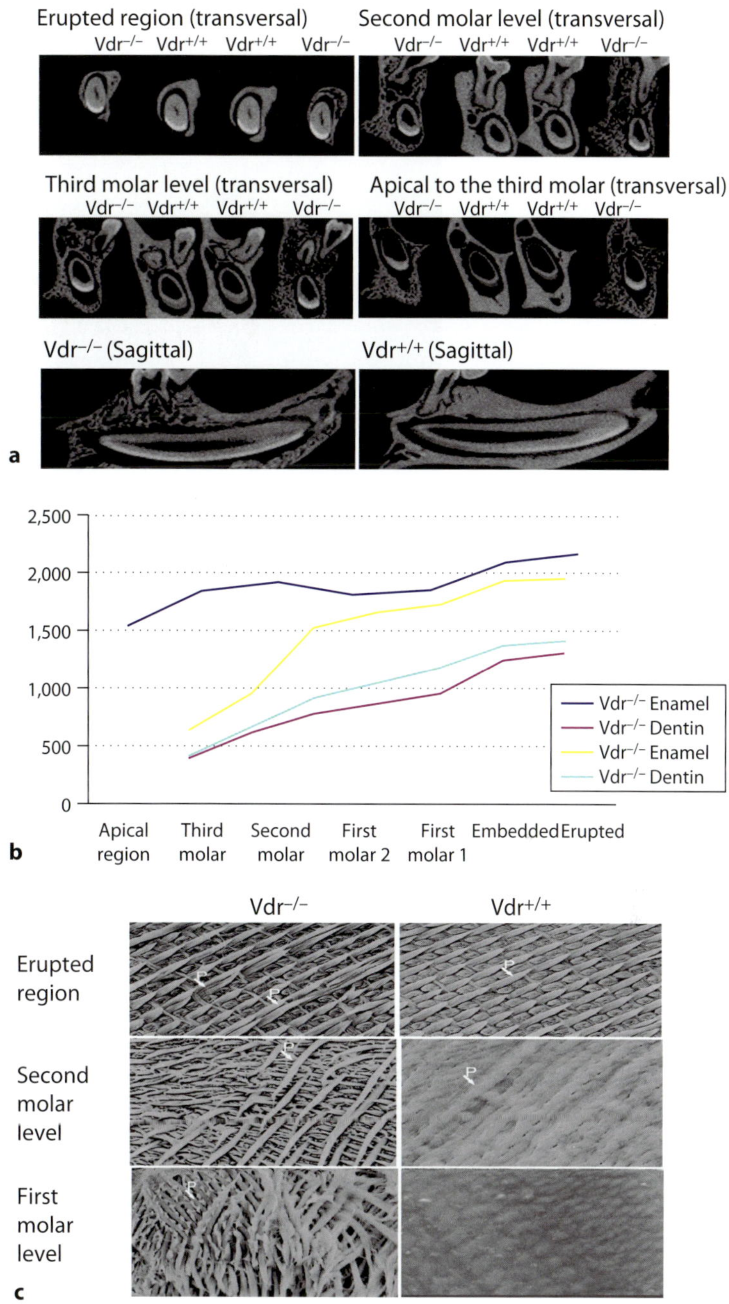

Fig. 2. Mineral deposition pattern of enamel and dentin observed in VDR wild-type (*vdr+/+*) and VDR knockout mice (*vdr–/–*). **a** Micro-CT images revealed that early enamel hypermineralization was observed apically to the second molar level in the *vdr–/–* mice. **b** Different distribution pattern of mineral deposition of enamel and dentin along the axis of incisor in *vdr+/+* and *vdr–/–* mice. A much divergent pattern between enamel and dentin was observed in *vdr–/–* group. **c** SEM image of enamel in different levels of *vdr+/+* and *vdr–/–* mouse incisor. Early enamel maturation was observed in the apical regions of *vdr–/–* group.

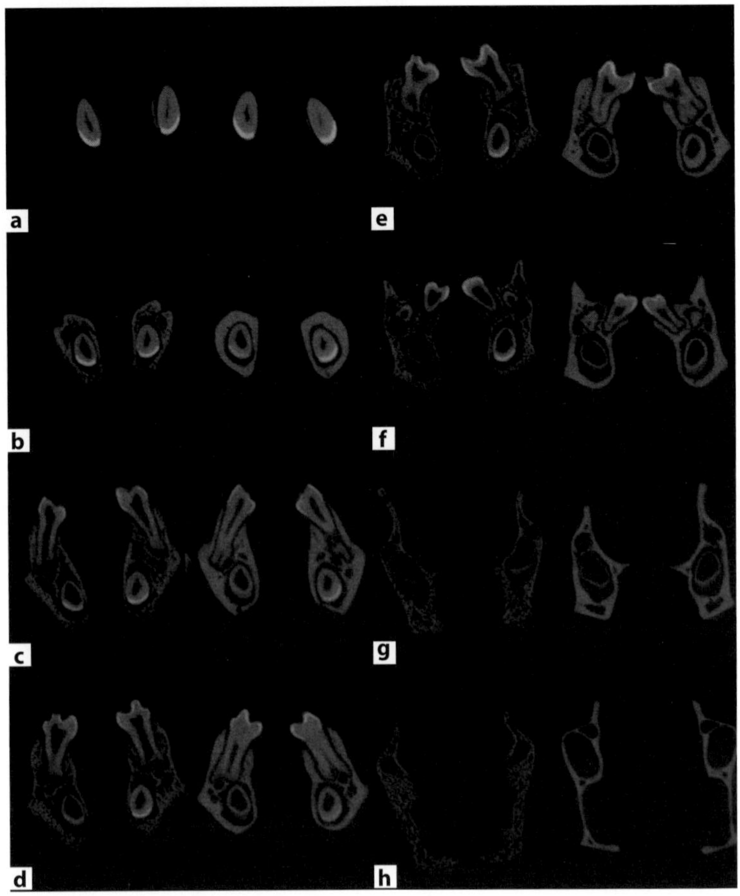

Fig. 3. Micro-CT images at different levels of the incisors of VDR wild-type (*vdr+/+*) and VDR knockout mice (*vdr–/–*) with accelerated eruption and normal eruption The maturation zone of the enamel was observed to move anteriorly in both VDR–/– and VDR+/+ groups. The VDR+/+ group showed typical enamel hypomineralization in the disoccluded incisor. In contrast, enamel mineralization was observed much faster in the disoccluded incisor of VDR–/– group than was seen in VDR+/+ group. **a** Erupted region. **b** Embedded mature region. **c** Region underlying the medial first molar. **d** Region underlying the distal first molar. **e** Region underlying the second molar. **c** Region underlying the third molar. **g** Apical region. **h** Ramus region.

mandible was disoccluded; the contralateral side was used as control. The incisor was shortened to the gingival margin every 2 days for 10 days, and then examined the following week. In our study, the maturation zone of the enamel was observed to move anteriorly in both VDR–/– and VDR+/+ groups. The VDR +/+ group showed typical enamel hypomineralization in the disoccluded incisor as reported previously [11, 12]. In contrast, we observed an acceleration of enamel mineralization in the disoccluded incisor of VDR–/– group (fig. 3). While delayed eruption may influence enamel mineralization as reported earlier [11–12], this was not observed in our studies. We conclude that the VDR deficiency may be a more important factor than delayed eruption to early mineralization.

Possible Explanations

Bone defects in VDR-deficient mice can be prevented by bypassing active calcium absorption with a diet containing high levels of lactose and calcium [13]. It has also been reported that no skeletal defect was observed in the osteoblast-specific conditional VDRKO mouse model [13]. These observations indicate that the principle action of the VDR in skeletal growth, maturation, and remodeling, is related to intestinal calcium absorption [2]. The skeletal consequence of VDR deficiency is a result of impaired intestinal calcium absorption and/or the resultant secondary hyperparathyroidism. Thus, vitamin D regulates bone mineralization primarily as an endocrine factor [14]. Osteoporosis in part develops as the major bone defect when vitamin D is deficient. Three main mechanisms underlie the pathogenesis of osteoporosis: insufficient mass during growth; excessive bone resorption; and inadequate new bone formation [15]. Recent research has found that mice containing a deleted VDR gene cannot produce RANKL in response to $1,25(OH)_2D_3$ and cannot support osteoclastogenesis [16]. Therefore, the osteoporosis-like phenotype in VDR deficient mice may not result from the bone remodeling and bone resorption, but most likely results from insufficient mineral deposition.

Enamel, dentin, and bone are mineralized tissues that develop through matrix-mediated mineralization process; however, the origin, composition and remodeling of these tissues are different [17]. Multiple studies in mice have reported dentin hypomineralization when vitamin D pathway disturbances exist [6–8]. Since dentin shares similar mechanism of mineralization to bone [18], the assumption is to attribute the dentin hypomineralization to inadequate levels of calcium and phosphorus in plasma. Additional research has revealed that the effects of $1,25(OH)_2D_3$ on bone mineralization not only maintains mineral homeostasis but also stimulates the formation of osteoclasts and promotes production of calcium binding proteins such as osteocalcin and osteopontin by osteoblasts. The synthesis of these proteins is positively associated with new bone formation and mineralization [2].

Although the mechanisms are unclear, VDR deficiency influences odontoblasts. VDR is expressed in odontoblasts, and odontoblasts and ameloblasts are the target cells for vitamin D function [19]. Recent investigations suggested that $1,25(OH)_2D_3$ functions to upregulate VDR, which in turn induces structural gene products, including calcium-binding proteins and several extracellular matrix proteins, such as dentin sialoglycoproteins and dentin phosphoproteins which are important for dentin formation [7, 20]. Therefore, the dentin hypomineralization, especially the translucent areas observed by micro-CT in our studies, may be caused by the abnormal structure and function of these extracellular matrix proteins in dentinogenesis. While it is clear that the dentin hypomineralization follows the trend of hypocalcemia, we believe that the calcium levels in plasma play major roles in dentin mineralization.

As the early enamel hypermineralization is independent of extracellular hypocalcemia, we propose that vitamin D deficiency mainly functions on enamel mineralization locally. The direct action of vitamin D on the target cells usually functions through genomic and nongenomic pathways and the early enamel mineralization which we have observed may be secondary to the absence of hormone-dependent VDR actions in ameloblasts. The genomic actions of vitamin D occur by binding the ligand to the VDR, which then heterodimerizes with the retinoid

X receptor (RXR). This dimer binds to vitamin D response elements (VDREs) on target genes. Unliganded VDR may repress a subset of target genes in a manner similar to other nuclear receptors through corepressor interaction [21]. Such VDR-RXR-repressor genes could be negatively involved in regulation of cell activities. Recent studies have revealed this inhibitory role of VDR on bone formation [1]. When mineral homeostasis was normalized in VDR knockout mice by the rescued diet, their bone showed high rates of bone formation in vivo [3]. In addition, cultured osteoblast precursors lacking VDR also had higher osteogenic potential [22].

The toxicity of $1,25(OH)_2D_3$ is also worth mentioning. Since the lack of binding with VDR, accumulated high levels of $1, 25(OH)_2D_3$ or its metabolites may interact with an alternative receptor in non-genomic pathway [2]. $1,25(OH)_2D_3$ membrane-associated, rapid response steroid-binding protein ($1,25D_3$-[MARRS]) on the surface of cells is not only involved in the cellular handling of $1, 25(OH)_2D_3$, but also participates in rapid changes in intracellular calcium concentrations, and alterations in membrane phospholipids metabolism [23]. In addition, the activity of the $1,25D_3$-[MARRS] protein is also integrally linked with the rapid actions of $1, 25(OH)_2D_3$ in the initiation of phosphate transport [24]. Supportive evidence has been reported in multiple studies [8, 24, 25]; however, the expression of $1,25D_3$-[MARRS] in VDR–/– ameloblasts and its potential to regulate mineral deposition is still under investigation.

Conclusions

VDR deficiency results in different distribution patterns of enamel and dentin mineral deposition in a continuously erupting mouse incisor; with a reduction in dentin mineralization and early enamel hypermineralization. Thus, vitamin D probably plays different roles in enamel and dentin mineralization. Vitamin D appears to regulate dentin mineralization indirectly, while having a direct influence on enamel mineralization.

References

1 Demay M: Mechanism of vitamin D receptor action. Ann NY Acad Sci 2006;1068:204–213.

2 Demay M, Sabbagh Y, Carpenter T: Calcium and vitamin D: what is known about the effects on growing bone. Pediatrics 2007;119:S141–S144.

3 Fleet JC: Molecular regulation of calcium and bone metabolism through the vitamin D receptor. J Musculoskelet Neuronal Interact 20;6:336–337.

4 Christakos S, Dhawan P, Benn B, et al: Vitamin D: molecular mechanism of action. Ann NY Acad Sci 2007;1116:340–348.

5 Ranggård L, and Norén J: Effect of hypocalcemic state on enamel formation in rat maxillary incisors. Scand J Dent Res 1994;102:249–253.

6 Giunta JL: Dental changes in hypervitaminosis D. Oral Surg Oral Med Oral Pathol Oral Radiol Endo 1998;85:410–413.

7 Lézot F, Descroix V, Mesbah M, et al: Cross-talk between Msx/Dlx homeobox genes and vitamin D during tooth mineralization. Connect Tissue Res 2002;43:509–514.

8 Lézot F, Descroix V, Hotton D, et al: Vitamin D and tissue non-specific alkaline phosphatase in dental cells. Eur J Oral Sci 2006;114:178–182.

9 Li Y, Pirro A, Amling M, et al: Targeted ablation of the vitamin D receptor: an animal model of vitamin D-dependent rickets type II with alopecia. Proc Natl Acad Sci 1997;94:9831–9835.

10 Smith C, Nanci A: A method for sampling the stages of amelogenesis on mandibular rat incisors using the molars as a reference for dissection. Anat Rec 1989;225:257–266.

11 Robinson C, Kirkham J, Nutman CA: Relationship between enamel formation and eruption rate in rat mandibular incisors. Cell Tissue Res 1988;254:655–658.

12 Risnes S, Septier D, Goldberg M: Accelerated eruption of rat lower incisor. Relationship between impeded and unimpeded eruption rates, rate of attrition, tooth length, and production of dentin and enamel. Connect Tissue Res 1995:32:183–189.

13 Takeyama K, Yamamoto Y, Kato S: VDR knockout mice and bone mineralization disorders. Clin Calcium 2007;17:1560–1566.

14 Suda T, Ueno Y, Fujii K, et al: Vitamin D and bone. J Cell Biochem 2003;88:259–266.
15 Raisz GL: Pathogenesis of osteoporosis: concepts, conflicts, and prospects. J Clin Invest 2005;115:3318–3325.
16 Takeda S, Yoshizawa T, Nagai Y, et al: Stimulation of osteoclast formation by 1,25-dihydroxyvitamin D requires its binding to vitamin D receptor (VDR) in osteoblastic cells: studies using VDR knockout mice. Endocrinology 1999;140:1005–1008.
17 Simmer J, Hu J: Expression, structure, and function of enamel proteinases. Connect Tissue Res 2002;43:441–449.
18 Qin C, Baba O, Butler WT: Post-translational modifications of sibling proteins and their roles in osteogenesis and dentinogenesis. Crit Rev Oral Biol Med 2004;15:126–136.
19 Berdal A, Papagerakis P, Hotton D, et al: Ameloblasts and odontoblasts, target-cells for 1,25-dihydroxyvitamin D_3: a review. Int J Dev Biol 1995;39:257–262.
20 Papagerakis P, MacDougall M, Berdal A: Differential epithelial and mesenchymal regulation of tooth-specific matrix proteins expression by 1,25-dihydroxyvitamin D_3 in vivo. Connect Tissue Res 2002;43:372–375.
21 Hendy GN, Hruska KA, Mathew S, et al: New insights into mineral and skeletal regulation by active forms of vitamin D. Kidney Int 2006;69:218–223.
22 Sooy K, Sabbagh Y, Demay M: Osteoblasts lacking the vitamin D receptor display enhanced osteogenic potential in vitro. J Cell Biochem 2005;94:81–87.
23 Wali RK, Kong J, Sitrin MD, et al: Vitamin D receptor is not required for the rapid actions of 1,25-dihydroxyvitamin D_3 to increase intracellular calcium and activate protein kinase C in mouse osteoblasts. J Cell Biochem 2003;88:794–801.
24 Nemere I, Safford SE, Rohe B, et al: Identification and characterization of 1,25D3-membrane-associated rapid response, steroid (1,25D3-MARRS) binding protein. J Steroid Biochem Mol Biol 2004;89–90:281–285.
25 Teillaud C, Nemere I, Boukhobza F, et al: Modulation of 1alpha,25-dihydroxyvitamin D3-membrane associated, rapid response steroid binding protein expression in mouse odontoblasts by 1alpha,25-(OH)2D3. J Cell Biochem 2005;94:139–152.

Dr. Xueming Zhang
Departments of Prosthodontics, School of Dentistry, University of Alabama at Birmingham
Birmingham, AL 35294–0007 (USA)
Tel. +1 205 934 7823, Fax +1 205 975 6108, E-Mail zhangxm@uab.edu

Koppe T, Meyer G, Alt KW (eds): Comparative Dental Morphology.
Front Oral Biol. Basel, Karger, 2009, vol 13, pp 110–115

Patterns of Asymmetry in Primary Tooth Emergence of Australian Twins

S. Mihailidis · S.N. Woodroffe · T.E. Hughes · M.R. Bockmann · G.C. Townsend

Craniofacial Biology Research Group, School of Dentistry, The University of Adelaide, Adelaide, S.A., Australia

Abstract

Aims: This study is part of a larger investigation of genetic and environmental influences on primary tooth emergence in Australian twins. Our aims were to describe patterns of emergence asymmetry, including directional and fluctuating components (DA, FA), and to test for a genetic basis to observed asymmetry. **Methods:** The study sample consisted of 131 twin pairs. Using one randomly-selected twin from each pair, dental asymmetry was examined by analysing the number of days between emergence of antimeres (Δ), with dates of emergence provided through parental recording. Scatterplots were used for assessment of DA and FA, followed by paired t-tests to detect significant differences in mean Δ from zero (evidence of DA). FA was assessed by calculating means and variances of the absolute value of Δ. A range of intervals (0, 7, 14, 21, 28 days) was used to define symmetrical emergence of antimeres. **Results:** Although a trend in left-side advancement for tooth emergence was detected, this was not statistically significant. Relatively low levels of FA were noted throughout the primary dentition, with maxillary and mandibular lateral inisors displaying the highest values, but no evidence of a genetic influence on FA was noted. Around 50% of all antimeric pairs of primary teeth were found to emerge within 14 days of each other, although time differences of more than 50 days were noted in some cases. **Conclusion:** Studies of dental asymmetry provide insights into the biological basis of lateralisation in humans and the results can also assist clinicians to discriminate between normal and abnormal developmental patterns.

As part of a larger study of genetic and environmental influences on primary tooth emergence in Australian twins, this analysis was conducted to identify whether patterns of asymmetry for tooth emergence exist within the primary dentition and to also determine whether the nature and extent of asymmetry differs between sexes or zygosity groups. Although there have been many studies of asymmetry in tooth size, there have been relatively few that have focussed on tooth emergence, particularly in the primary dentition.

Asymmetry is said to be directional (DA) when one side regularly displays greater and/or earlier development than the other, reflecting an underlying biological mechanism designed to serve a particular functional purpose. Random, non-directional differences between sides, termed fluctuating asymmetry (FA), are thought to indicate the inability of an individual to buffer against developmental disturbances.

DA in primary teeth has been identified for crown size [1] and left-advanced emergence of primary teeth has also been reported [2]. Various patterns of FA have been reported in relation to arch and tooth groups [3–6]. Previous studies of the primary dentition have shown no association of asymmetry

All authors are associated with the International Collaborating Centre in Oro-facial Genetics and Development.

with gender [1, 2, 4] and significantly less asymmetry compared with the secondary dentition [7].

A recent study has shown strong genetic control for emergence timing of the primary incisors [8]. What remains less clear is the degree of genetic and/or environmental influence on emergence asymmetry and the nature of the biological basis to body lateralisation. Are DA and FA in dental structures genetically determined? Results have varied from less asymmetry in twins compared with singletons [9], to no significant difference between the two groups [1, 10]. Furthermore, although examples of mirror-imaging have been observed in the dentitions of twins [11], there have been no detailed analyses of the frequency of this phenomenon in different zygosity groups.

If patterns of asymmetry within the dentition can be understood in relation to underlying biological mechanisms, our understanding of early human development, and dental development in particular, will be increased.

Hence, the aims of this study were to:

- Describe patterns of asymmetry in primary tooth emergence
- Determine the presence and extent of DA and FA
- Test for associations of asymmetry with gender and between MZ and DZ co-twins
- Describe the trends in asymmetry when different time intervals for emergence between antimeres were defined

Materials and Methods

The study sample consisted of 131 twin pairs including 59 pairs of monozygotic (MZ), 40 pairs of dizygotic same-sex (DZ_{SS}) and 32 pairs of dizygotic opposite-sex (DZ_{OS}) twins.

Using one randomly selected twin from each pair, dental asymmetry was examined by analysing the number of days between emergence of antimeres (Δ). Scatterplots of emergence times of antimeric pairs provided an initial assessment of the presence of DA and FA, followed by more formal testing. Firstly, DA was quantified by testing whether the mean Δ differed from zero (paired t tests, alpha = 0.05). The presence and extent of FA was then quantified by calculating means and variances of the absolute differences in emergence times of antimeres $|\Delta|$. Comparisons of $|\Delta|$ values were made between males and females to determine the influence of gender on FA. Twin comparisons were also conducted to examine whether there were associations between MZ and DZ co-twins for asymmetry. Several time intervals (0, 7, 14, 21, 28 days) were used to define symmetrical versus asymmetrical antimeric emergence and comparisons were made between different tooth types, e.g. incisors, canines and molars. For example, when using a 14-day interval, antimeres were considered to emerge symmetrically if they emerged within 14 days of each other. However, if they emerged more than 14 days apart, they were categorised as displaying asymmetrical emergence with either left- or right-side advancement.

Results

Scatterplots (not presented) demonstrated a strong correlation between antimeres for timing of emergence ($0.95<r<1.0$) and, relative to the hypothetical line of symmetry, a small degree of FA for all primary teeth. The mandibular lateral incisors, followed by the maxillary lateral incisors, exhibited most scatter, reflecting greater FA compared with other tooth groups. However, there was no evidence of DA and this was confirmed using paired t-tests. Furthermore, there was no evidence of difference in asymmetry between the sexes.

The presence ($|\Delta|$) and extent (s^2) of FA is shown in table 1. The second molars have been excluded due to small sample sizes. The absolute means $|\Delta|$ in number of days between emergence of both antimeres for maxillary and mandibular lateral incisors were almost double those for other antimeric pairs, while the variances (s^2) were more than triple. The difference in emergence times between lateral incisors was 24–28 days whereas all other antimeric pairs averaged 10–17 days. This finding indicates that lateral incisors, particularly mandibular, exhibit relatively greater FA compared with other primary teeth.

Table 1. Magnitude of FA in the primary dentition of Australian twins

Antimeric pairs		Mean $\|\Delta\|$	s^2
Maxilla	di1	12	288
	di2	24	1,831
	dc	12	436
	dm1	10	300
Mandible	di1	13	544
	di2	28	1,664
	dc	13	429
	dm1	17	616

$|\Delta|$ denotes the absolute mean difference in emergence of antimeres (in days).
s^2 denotes the variance of difference in emergence of antimeres (in days^2).

Scatterplots comparing the degree of correlation between MZ and DZ twin pairs were used to ascertain whether observed FA was genetically linked. Correlations greater than zero and relatively higher between MZ co-twins than DZ co-twins would indicate a genetic influence. However, there was no evidence of a genetic basis to FA as points were randomly distributed (fig. 1).

When emergence of antimeres was examined for evidence of patterning over increasing intervals of time, no significant trend could be detected between tooth types (fig. 2). However, despite the lack of a statistically significant finding for DA, left-side advancement was observed for the maxillary canine and first molar and the mandibular central incisor, canine and first molar when increasing intervals of time were used to define symmetrical emergence between antimeres (fig. 2). With the exception of maxillary canines, few antimeres emerged on the same day. Once symmetrical emergence was defined by a longer time period (7, 14, 21 and 28 days), as expected, increasingly more antimeric pairs were considered to emerge symmetrically. For example, 50% of all antimeres emerged symmetrically when a 14-day time period was applied. The lateral incisors were exceptions, being consistently more likely to emerge asymmetrically for all time intervals (fig. 2).

Discussion

Low levels of FA for tooth emergence were demonstrated within the primary dentition of Australian twins, with increased FA for the lateral incisors. While DA was not statistically confirmed, some left-side dominance was shown when teeth were examined at increasing intervals of time, a trend consistent with earlier studies for both the primary [2] and secondary [12, 13] dentitions. Where DA was demonstrated in an earlier study of the primary dentition, the majority of teeth in males showed directionality but only the mandibular second primary molars demonstrated significance in females [2]. Our study did not find any significant sex difference for either DA or FA, consistent with other studies reporting a lack of significant sex effect on asymmetry in emergence for both primary [1, 4, 14] and secondary [15] teeth.

Heikkinen et al. [12, 13] found a difference in emergence timing of 1–6 months to be normal for permanent antimeres but our results indicate a smaller degree of FA in the primary dentition with most antimeres emerging within 2–3 weeks of each other. Similar trends of smaller FA in the primary dentition compared with the secondary dentition have been reported in relation to tooth size [16]. It has been suggested that the controlled intra-uterine environment in which primary tooth crowns develop may provide a greater buffering effect against developmental disturbances compared to that of the permanent dentition [16, 17]. The relatively short period of

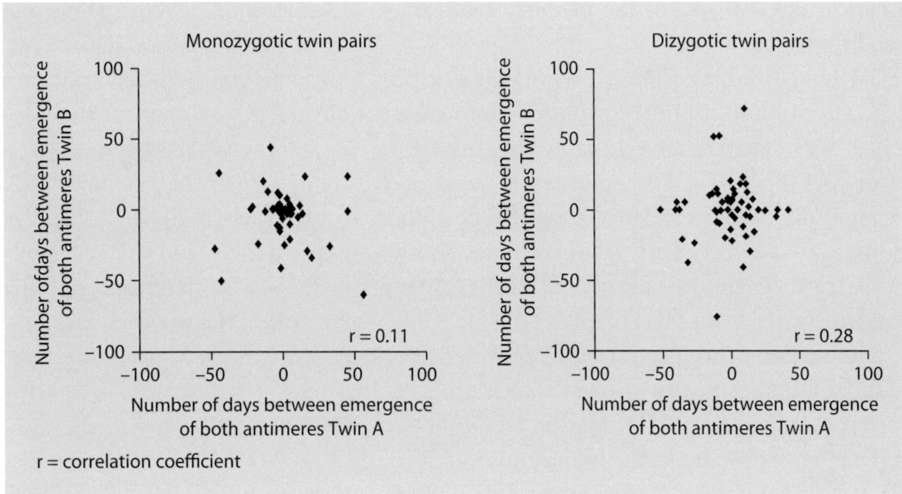

Fig. 1. Scatterplots showing differences in antimeric tooth emergence of primary mandibular lateral incisors between monozygotic and dizygotic twin pairs. r = correlation coefficient.

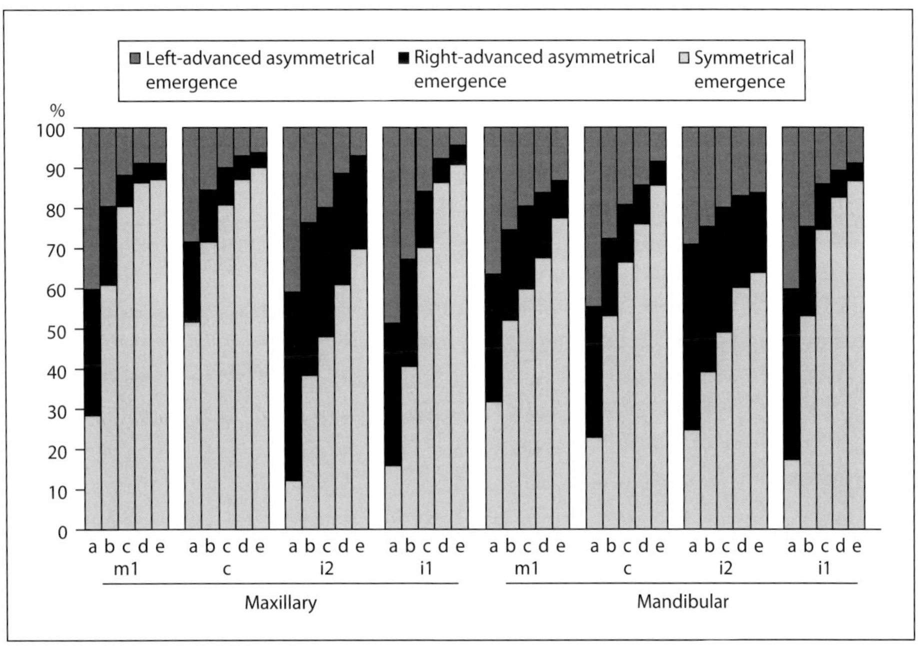

Fig. 2. Frequencies of occurrence of symmetrical and asymmetrical (left-advanced, right-advanced) emergence of antimeres in the primary dentition based on increasing intervals of time difference: (**a**) 0, (**b**) 7, (**c**) 14, (**d**) 21 and (**e**) 28 days.

time for development of the primary teeth may also be a factor.

Twin studies have shown a strong genetic basis to variation in timing of primary tooth emergence, with heritability estimates ranging from 78 to 96% [8, 18, 19]. Our analyses showed that correlations for FA between MZ and DZ co-twins did not differ significantly from zero for all antimeric teeth. Therefore, a genetic basis to FA was not detectable at this level of analysis.

The degree of asymmetry and prevalence of dental anomalies in the primary dentition is reported to be higher in the anterior region of the oral cavity, especially in the mandibular arch [6, 20]. Possible explanations for this finding include patterns in innervation [21], site and timing of embryological development [22] and evolutionary trends toward a reduction in jaw size [23]. It is hypothesised that, as innervation occurs in the dental follicle which is essential for eruption, innervation must be related to emergence/eruption. It is suggested that regions with more separated pathways of nerve supply may display less variation, for example maxillary compared to mandibular teeth [21]. This may explain why lower lateral incisors, situated close to a site of nerve branching (incisive and mental branches of the inferior alveolar nerve), might be more variable. Increased developmental vulnerability of the maxillary lateral incisors has been linked to the process of fusion between the medial and lateral nasal processes which occurs at the site of tooth formation [22]. While this is plausible in explaining maxillary lateral incisor variation, reasons for lower lateral incisor variation remain less clear. Similarly, variation in the permanent maxillary lateral incisor has been linked to an evolutionary reduction of the premaxilla [23]. Although this finding relates to permanent incisors, it may also explain increased variability observed for primary incisor tooth groups, maxillary and mandibular, compared to other primary tooth groups.

Conclusion

Our study has shown evidence of FA in primary tooth emergence, particularly for lateral incisors, and a tendency for left-side advancement for some teeth. Half of all primary antimeric pairs emerged within 2 weeks of each other, a finding that should assist clinicians to discriminate between normal and abnormal developmental patterns. No evidence was found for a genetic contribution to FA in tooth emergence.

Acknowledgements

The support of the NHMRC of Australia, ADRF and the Dental Board of South Australia is gratefully acknowledged. We thank the twins and their families who agreed to participate, the Australian Twin Registry and Sandy Pinkerton for her assistance.

References

1 Townsend GC, Dempsey P, Richards LC: Asymmetry in the deciduous dentition: fluctuating and directional components. Perspect Hum Biol 1999;4:45–52.
2 Lysell L, Magnusson B, Thilander B: Time and order of eruption of the primary teeth. Odont Rev 1962;13:217–234.
3 Black TK: Fluctuating asymmetry in the deciduous dentition. J Dent Res 1980;59:725.
4 Townsend GC: Fluctuating asymmetry in the deciduous dentition of Australian Aboriginals. J Dent Res 1981;60:1849–1857.
5 Farmer V, Townsend GC: Crown size variability in the deciduous dentition of South Australian children. Am J Hum Biol 1993;5:681–690.
6 Liversidge HM, Molleson TI: Deciduous tooth size and morphogenetic fields in children from Christ Church, Spitalfields. Arch Oral Biol 1999;44:7–13.
7 Garn SM, Bailey SM: The symmetrical nature of bilateral asymmetry in deciduous and permanent teeth. J Dent Res 1977;56:1422.

8 Hughes TE, Bockmann MR, Seow K, Gotjamanos T, Gully N, Richards LC, Townsend GC: Strong genetic control of emergence of human primary incisors. J Dent Res 2007;86:1160–1165.
9 Boklage CE: Developmental differences between singletons and twins in distributions of dental diameter asymmetries. Am J Phys Anthropol 1987;74:319–331.
10 Staley RN, Green LJ: Bilateral asymmetry in tooth cusp occurrence in human monozygotic twins, dizygotic twins, and nontwins. J Dent Res 1971;50:83–89.
11 Townsend GC, Richards LC, Brown T: Mirror imaging in the dentitions of twins – what is the biological basis?; in Brown T, Molnar S (eds): Craniofacial Variation in Pacific Populations. Adelaide, Anthropology and Genetics Laboratory, University of Adelaide, 1992, pp 67–78.
12 Heikkinen T, Alvesalo L, Osborne RH, Tienari J, Virtanen JI: The effects of sex and race on the symmetry of the permanent tooth clinical eruption; in Mayhall JT, Heikkinen T (eds): Dental Morphology 1998. Oulu, Oulu University Press, 1999, pp 473–482.
13 Heikkinen T, Alvesalo L, Osborne RH, Tienari J: Tooth eruption symmetry in functional lateralities. Arch Oral Biol 2001;46:609–617.
14 Townsend GC, Garcia-Gordoy F: Fluctuating asymmetry in the deciduous dentition of Dominican mulatto children. Arch Oral Biol 1984;29:483–486.
15 Sobhi P, Mihailidis S, Rogers JR, Hughes TE, Townsend GC: Asymmetrical eruption of permanent teeth in Australian Aborigines. Dent Anthropol 2007;20:33–41.
16 Guatelli-Steinberg D, Scuilli PW, Edgar HJ: Dental fluctuating asymmetry in the Gullah: tests of hypotheses regarding developmental stability in deciduous vs. permanent and male vs. female teeth. Am J Phys Anthropol 2006;129:427–434.
17 Cook DC, Buikstra JE: Health and differential survival in prehistoric populations: prenatal dental defects. Am J Phys Anthropol 1979;51:649–664.
18 Hatton ME: A measure of the effects of heredity and environment on eruption of the deciduous teeth. J Dent Res 1955;34:397–401.
19 Townsend GC, Richards L, Messer L, Hughes T, Pinkerton S, Seow K, Gotjamanos T, Gully T, Bockmann M: Genetic and environmental influences on dentofacial structures and oral health: studies of Australian twins and their families. Twin Res Hum Genet 2006;9:727–732.
20 Yonezu T, Hayashi Y, Sasaki J, Machida Y: Prevalence of congenital dental anomalies of the deciduous dentition in Japanese children. Bull Tokyo Dent Coll 1997;38:27–32.
21 Parner ET, Heidmann JM, Kjaer I, Vaeth M, Poulsen S: Biological interpretation of the correlation of emergence times of permanent teeth. J Dent Res 2002;81:451–454.
22 Hovorakova M, Lesot H, Peterkova R, Peterka M: Origin of the deciduous upper lateral incisor and its clinical aspects. J Dent Res 2006;85:167–171.
23 Mizoguchi Y: Size stability of the maxillary lateral deciduous incisor. Anthropol Sci 2002;110:365–388.

Dr. Suzanna Mihailidis
Craniofacial Biology Research Group, School of Dentistry, The University of Adelaide
Adelaide SA 5005 (Australia)
Tel. +61 8 8303 6788, Fax +61 8 8303 3444, E-Mail suzanna.mihailidis@adelaide.edu.au

Koppe T, Meyer G, Alt KW (eds): Comparative Dental Morphology.
Front Oral Biol. Basel, Karger, 2009, vol 13, pp 116–120

Rates of Enamel Formation in Human Deciduous Teeth

Wendy Birch · M. Christopher Dean

Department of Cell and Developmental Biology, University College London, London, UK

Abstract

The aim of this study was to document rates of enamel formation in deciduous teeth. Little is known about rates of deciduous enamel formation compared to permanent enamel. In permanent teeth, rates vary between 2.5 μm per day at the EDJ to 6.5 μm per day at the enamel surface. Longitudinal ground sections of twenty mandibular deciduous teeth (4 of each tooth type) made through the crown in the buccolingual plane were selected that showed clearly visible daily enamel cross striations using transmitted polarised light microscopy. Ten average measurements, each one made across five daily increments, were recorded within 100-μm thick zones defined between the EDJ and the enamel surface on each tooth section. This procedure was repeated in occlusal, lateral and cervical regions of each tooth. Overall, daily rates varied less than in permanent teeth and did not show the very low rates at the EDJ or the very high rates often found in the outer enamel of permanent teeth. In deciduous enamel, rates varied between 2.5 and 4.5 μm throughout, but often showed a marked reduction in the zone immediately following the neonatal line or other accentuated markings usually associated with stressful events. A catch-up phase usually followed these events during which rates recovered. These data provide clear evidence of enamel hypoplasia associated with both the birth process and other events that cause stress in perinatal life.

Cross-striations, or alternating varicosities and constrictions, along enamel prisms have long been accepted to be daily incremental markings in both permanent teeth and deciduous teeth [1–7]. It follows that measurements of the spacing of enamel cross striations can be used to estimate the daily linear secretion rates of enamel matrix by ameloblasts. However, very few histological studies of human deciduous teeth exist where rates of enamel formation have been estimated systematically. However, rates of enamel formation in human permanent teeth are much better documented and range from 2.5 to 6.5 μm/day [8, 9]. The aim of this study was to document rates of enamel formation in deciduous teeth in three regions of different deciduous tooth types. All deciduous teeth begin their formation before birth and contain a neonatal line (NNL) that divides enamel formed prenatally from enamel formed postnatally. A further aim of this study was to record rates of enamel formation across the boundary from prenatal to postnatal in the cuspal, lateral and cervical regions of the crowns of each deciduous tooth type and describe any changes.

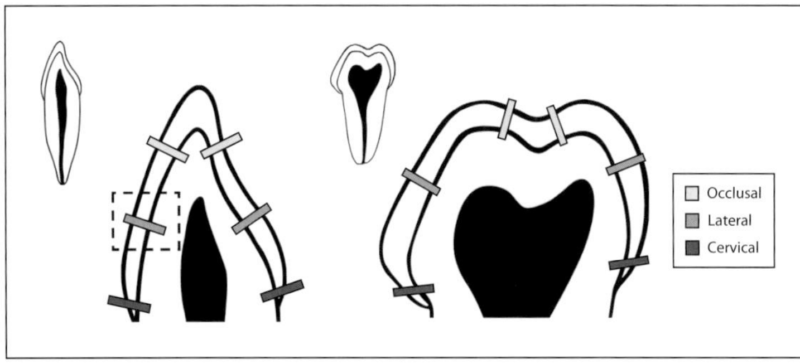

Fig. 1. Diagrams of longitudinal sections of a deciduous canine and molar showing how each section was divided into cervical, lateral and occlusal regions. The lateral boxed area of the canine section represents the region shown in figure 2a.

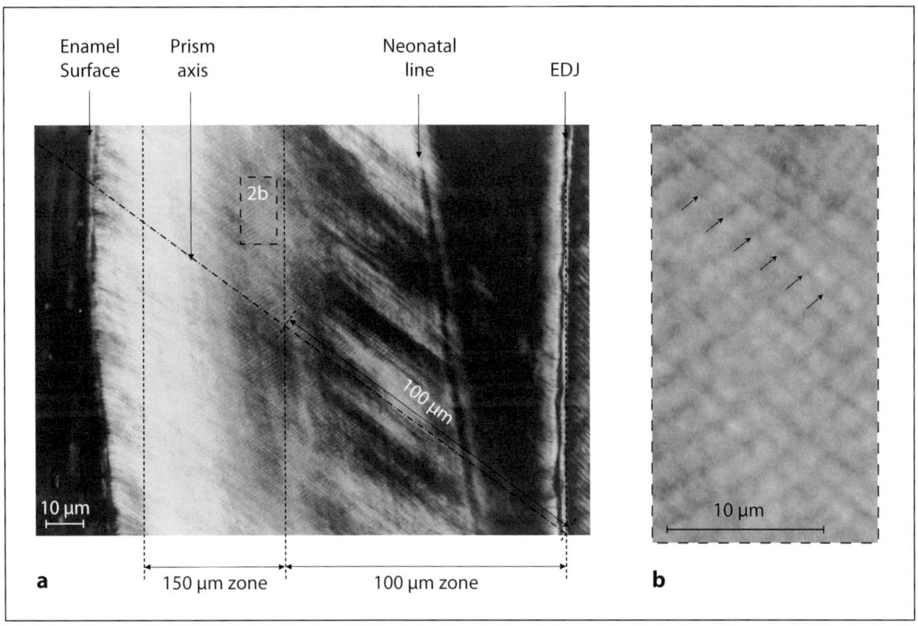

Fig. 2. a The lateral region of the labial aspect of a deciduous canine. The enamel-dentine junction (EDJ), neonatal line, enamel surface and prism path axis are indicated. The distance along the prism path axis from the EDJ was divided into 100-μm zones parallel to the EDJ. **b** Higher power portion of **a**. An example of 5 days of enamel growth marked by 6 arrows at cross-striations is shown.

Materials and Methods

Ground sections were made in the true longitudinal buccolingual axial plane of deciduous mandibular teeth. From a large sample, 20 ground sections were chosen that showed clearly visible daily enamel cross striations in transmitted polarised light that could be seen along prism paths running between the EDJ and the enamel surface. Four ground sections of each tooth type were chosen. Photomontages were constructed of the lingual and buccal aspects of each section. Each aspect of the crown was divided into occlusal, lateral and cervical regions (fig. 1). Digital calipers accurate to 0.01 μm were used to make all measurements on the montages. A magnification factor and a scale for the montages were calculated from images of a 1-mm graticule scored with 10-μm increments made using the same ×25 microscope objective lens. All measurements made on the montages were rounded up to the nearest 0.1 μm. Zones of enamel 100 μm thick were measured along continuous groups of

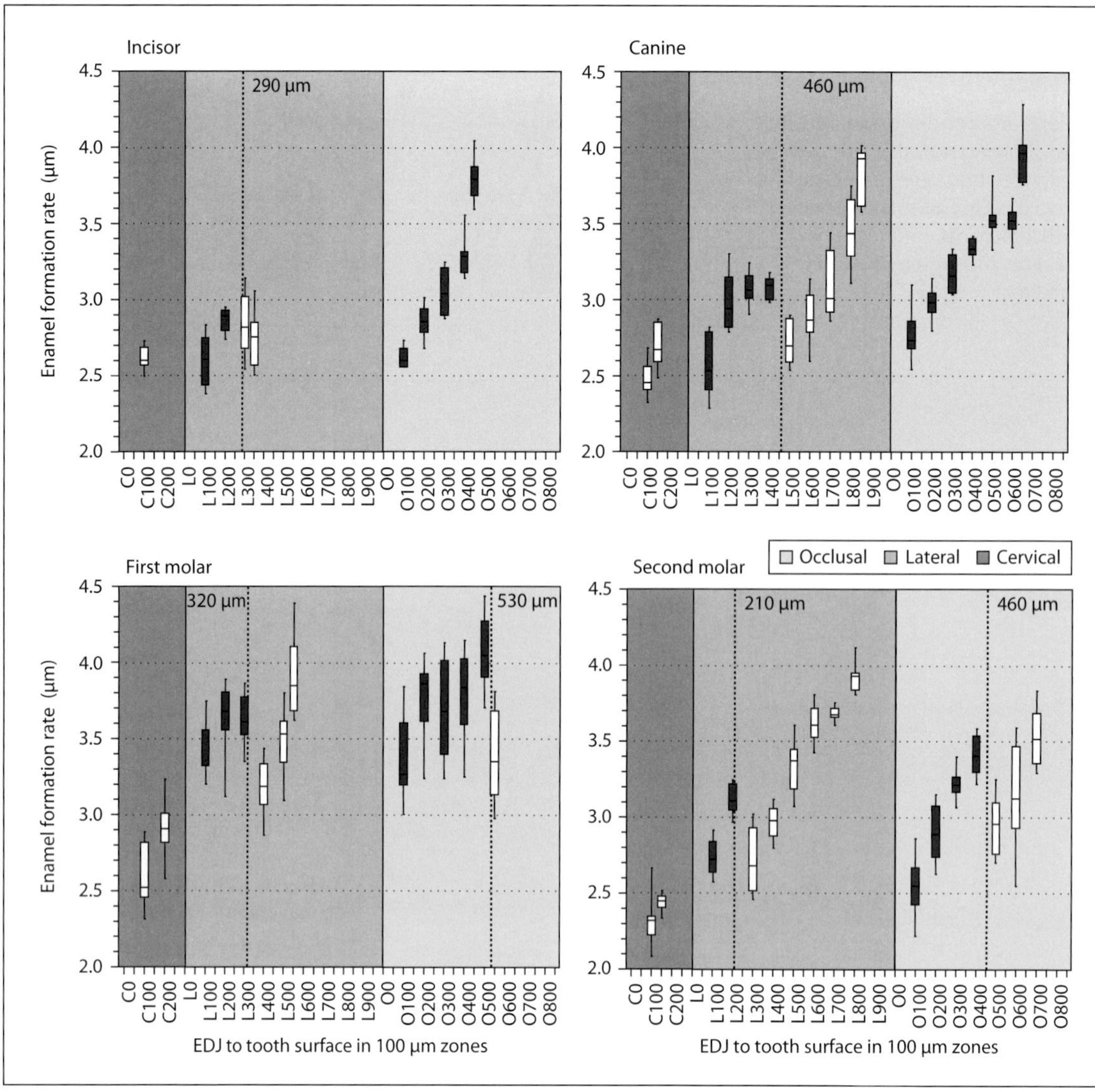

Fig. 3. Graphs of deciduous enamel formation rates in four deciduous mandibular teeth. Plots for a second deciduous incisor and canine appear on the top and those for a first and second molar beneath. Cervical, lateral and occlusal regions are denoted by different shades of grey background. The vertical black dashed line indicates the position of the NNL with respect to the EDJ. Its distance from the EDJ is indicated on the plots for each tooth type. Note that cervical or occlusal enamel in some teeth may not contain a NNL. Individual box plots for prenatal enamel are filled dark-grey and for postnatal enamel filled white. Each box plot is for ten groups of six cross-striations in every 100-µm zone of enamel measured from the EDJ. The horizontal lines show the 25th, 50th and 75th centiles, the whiskers indicate the 10th and 90th centiles. The distance from the EDJ has also been included.

prisms selected for clarity in each region of each aspect of each crown (fig. 2). If the enamel stopped significantly short of a 100 μm measurement at the surface, a 50-μm zone was included. In each 100-μm zone, the distance between a consecutive series of six cross striations (that represented 5 days continuous enamel growth) was measured and an average calculated. This procedure was repeated ten times at different locations throughout each 100 μm thick zone (fig. 3). This was repeated in each zone throughout the thickness of the enamel. The data for the ten average values in each zone were presented as box plots for each region of each aspect of each tooth type.

Results

In general, there was an increase in the rate of enamel formation from the EDJ to the surface enamel; the rate of formation was slightly slower in the cervical enamel compared to the cuspal enamel. Daily increments ranged between a minimum value of 2.5 μm at the EDJ and a maximum of 4.5 μm at the enamel surface occlusally and between 2.5 μm and 3.5 μm cervically.

Superimposed upon this pattern there was an almost universal sharp decrease in the rate of enamel formation in the 100-μm zone following either the NNL or certain other accentuated incremental markings. Generally, this decrease was of the order of up to 0.5 μm/day. Within the subsequent 100 μm zone, there was a catch-up phase where rates of enamel formation generally returned to their previous values. This reduction in enamel matrix secretion is clear evidence of enamel hypoplasia associated with a stress line in enamel. The hypoplastic phase and the catch-up phase generally occured within a single 100-μm zone of enamel thickness.

Discussion

Many people have studied the nature and position of the NNL in modern human teeth [10–14]. However, few have documented the rate of deciduous enamel formation [6, 7, 15, 16]. This, together with that of Macchiarelli et al. [16] is the first methodical study to demonstrate a hypoplastic reaction to a stress line (NNL) in deciduous teeth followed by a catch-up phase immediately afterwards. Despite the fact that teeth and brains are usually considered the most resilient to environmental stress, this study shows that at least deciduous enamel responds in the same way to stressful events as other tissues, for example, bone, cartilage, muscle and fat. It is well known that the physiological changes at birth are associated with loss of weight, autophagy, acidosis, etc. [17, 18]. Besides the physical appearance of the NNL in deciduous teeth, it is now clear that true enamel hypoplasia, either associated with or without a line [8], is an equally good measure of stress during deciduous crown formation as well as in permanent crown formation [8].

Conclusions

Deciduous tooth enamel forms at a more constant rate than permanent tooth enamel but gradients in secretion rates between the EDJ and surface and between cuspal and cervical regions still exist. Rates of enamel formation immediately after the NNL often drop by an average of 0.5 μm/day but then catch up within a 100-μm (roughly 1 month) zone.

Acknowledgements

We thank the following: Dr. Don Reid for the loan of deciduous tooth sections and we thank Dr. Helen Liversidge for her continuing advice and encouragement, Dr. Charles FitzGerald, Dr. Carsten Witzel, Dr. Huw Thomas and Dr. John Luckacs, for their helpful comments at the 14th ISDM 2008, and the referees for their helpful suggestions. This study is part of an ongoing research project funded by the Department of Cell and Developmental Biology at University College London.

References

1 Boyde A: Carbonate concentration, crystal centres, core dissolution, caries, cross striation, circadian rhythms and compositional contrast in the SEM. J Dent Res 1979;58b:981–983.

2 Boyde A: Enamel; in Berkovitz BKB, Boyde A, Frank RM, Höhling HJ, Moxham BJ, Nalbandian J, Tonge CH (eds): Teeth: Handbook of Microscopic Anatomy. New York, Springer, 1989, pp 309–473.

3 Bromage TG: Enamel incremental periodicity in the pig-tailed macaque; a polychrome fluorescent labelling study of dental hard tissues. Am J Phys Anthropol 1991;86:205–214.

4 Smith TM: Experimental determination of the periodicity of incremental features in enamel. J Anat 2006;208:99–113.

5 Antoine D, Hillson S, Dean MC: The developmental clock of dental enamel: A test for the periodicity of prism cross-striations and an evaluation of the likely sources of error in histological studies of this kind. J Anat 2009;214:45–55.

6 Schour I, Poncher H G: Rate of apposition of enamel and dentin measured by the effect of acute fluorosis. Am J Dis Child 1937;54:757–776.

7 Katzenberg MA, Oetelaar G, Oetelaar J, FitzGerald C, Yang D, Saunders SR: Identification of historical human skeletal remains: a case study using skeletal and dental age, history and DNA. Int J Osteoarchaeol 2005;15:61–72.

8 Mahoney P: Intraspecific variation in M1 enamel development in modern humans: implications for human evolution. J Hum Evol 2008;55:131–147.

9 Dean M C: A comparative study of cross striation spacings in cuspal enamel and of four methods of estimating the time taken to grow molar cuspal enamel in *Pan, Pongo* and *Homo*. J Hum Evol 1998;35:449–462.

10 Rushton MA: On the fine contour lines of the enamel of milk teeth. Dent Rec 1933;53:170–171.

11 Schour I: The neonatal line in the enamel and dentin of the human deciduous teeth and first permanent molar. JADA 1936;23:1946–1955.

12 Gustafson G, Gustafson AG: Microanatomy and histochemistry of enamel; in Miles AEW (ed): Structural and Chemical Organization of Teeth. London, Academic Press, 1967, vol 2, pp 75–134.

13 Eli I, Sarnat H, Talmi E: Effect of the birth process on the neonatal line in primary tooth enamel. Pediatr Dent 1989;11:3:220–223.

14 Skinner M, Dupras T: Variation in birth timing and location of the neonatal line in human enamel. J Forens Sci 1993;38;6:1383–1390.

15 Massler M, Schour I: The appositional life span of the enamel and dentin-forming cells. J Dent Res 1946;25:145–150.

16 Macchiarelli R, Bondioli L, Debénath, A, Mazurier A, Tournepiche J-F, Birch W, Dean C: How Neanderthal molar teeth grew. Nature 2006;444:748–751.

17 Okada, M: Hard tissues of animal body: Highly interesting details of Nippon studies in periodic patterns of hard tissues are described. Shanghai Evening Post Special Edition, Health, Recreation and Medical Progress 1943, pp 15–31.

18 Kuma A, Hatano M, Matsui M, Yamamoto A, Nakaya H, Yoshimori T, Ohsumi Y, Tokuhisa T, Mizushima N: The role of autophagy during the early neonatal starvation period. Nature 2004;432:1032–1036.

Wendy Birch, MSc
Department of Cell and Developmental Biology, University College London
Gower Street
London, WC1E 6BT (UK)
E-Mail w.birch@ucl.ac.uk

Koppe T, Meyer G, Alt KW (eds): Comparative Dental Morphology.
Front Oral Biol. Basel, Karger, 2009, vol 13, pp 121–127

Tooth Root and Craniomandibular Morphological Integration in the Common Chimpanzee (*Pan troglodytes*): Alternative Developmental Models for the Determinants of Root Length

Samuel N. Cobb · Hester Baverstock

Functional Morphology and Evolution Unit, Hull York Medical School, University of Hull, Hull, UK

Abstract

Background/Aims: Root length is strongly related to tooth stability but demonstrates considerable intraspecific variation. Previous studies have demonstrated an intraspecific relationship between root length and facial length in diverse mammalian taxa. These findings are indicative of plasticity in root length but with no clear developmental mechanism. This study aims to further these findings by identifying patterns of covariance between postcanine tooth root length and the whole integrated craniofacial skeleton in order to allow more refined hypotheses of the underlying developmental mechanisms to be proposed. **Methods:** 2D landmark coordinates were obtained from lateral radiographs of 27 adult *Pan troglodytes* skulls. The landmark configurations were divided into two blocks, one of craniofacial landmarks and another of landmarks related to mandibular tooth roots. Covariation of the two blocks was determined using partial least squares analysis. **Results:** The correlation coefficient between the first pair of singular warps is 0.76, highly significant ($p < 0.02$) and not sex related. Visualisation of this correlation shows a clear pattern of increasing root length variation along the tooth row with increasing facial height but not length. **Conclusions:** The findings support previous conclusions that tooth roots demonstrate plasticity during their development. A correlation between root length variation along the tooth row and facial height rather than length can be interpreted in the context of previous findings of maxillary and mandibular rotation and compensatory remodeling during development. It is therefore proposed that the observed root length plasticity is due to variation in the eruptive distance associated with compensatory jaw rotation during development.

While considerable effort has been expended on understanding the relationship between craniodental morphology and diet, the adaptive significance of root length has received relatively little attention. Despite this a limited number of studies demonstrate that tooth root surface area, and in turn length, undoubtedly reflect the ability of a tooth to resist occlusal loads experienced during food processing [1, 2]. Comparative studies on platyrrhine species have demonstrated that taxa with diets which require higher occlusal forces have tooth roots with significantly larger relative surface areas [2]. There is also evidence to suggest that variation in the magnitude of occlusal forces that can be generated at different bite points in the mouth of a single species is also reflected in tooth root form [3]. Although

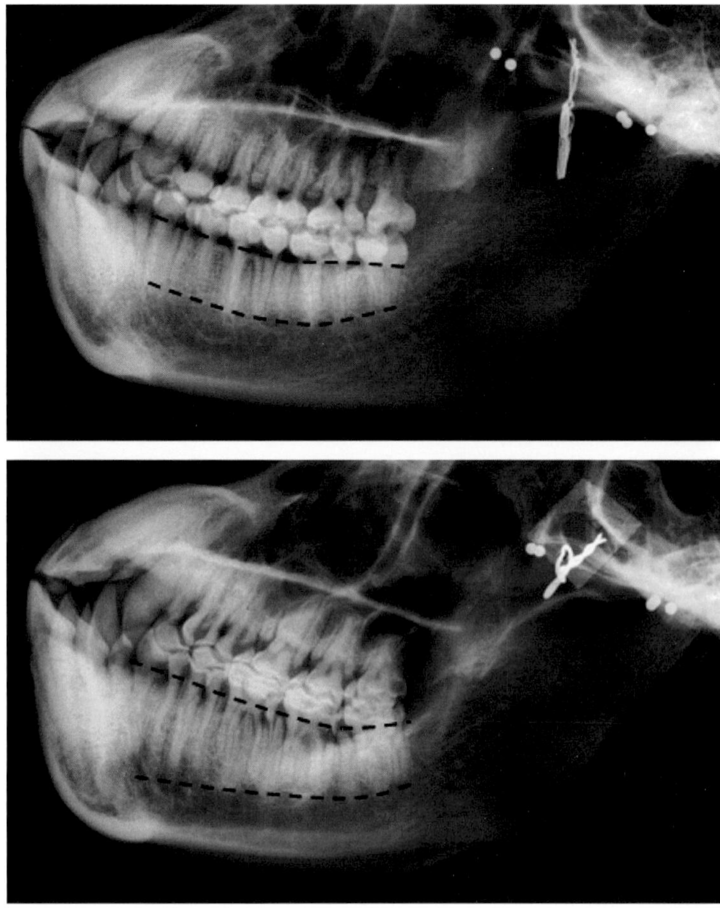

Fig. 1. Variation in root length along the tooth row in female *P. troglodytes*. The dotted lines denote the length of the roots.

tooth root length is closely related to tooth stability, root length demonstrates considerable intraspecific variation [4, 5]. Indeed, preliminary observations in *Pan troglodytes* show a high degree of variation not just in root length of the same teeth between individuals but also in the pattern of root length variation between tooth types (fig. 1).

Previous studies have noted the high degree of variation in the pattern of root length around the mouth in a number of species and have sought to investigate its developmental basis. Studies on rats and baboons with experimentally shortened faces [6, 7] and comparisons of long and short faced dog breeds [8] have demonstrated the plasticity of root length during development and a relationship between facial morphology and root length. While root length was not measured directly in the study [8] (root length was inferred from crown measurements and whole tooth weights), a gradient in root size reduction was found from the first to third molar, thus demonstrating a pattern of variation in root length of tooth types. These experimental studies demonstrate both a degree of plasticity in tooth root length during development and that the facial skeleton and tooth roots are developmentally integrated. However, the exact nature

Table 1. Landmark definitions

Craniofacial block landmarks
Superior most point on orbital margin
Inferior most point on orbital margin
Rhinion
Superior limit of premaxilla
Alveolare
Inferior point of incisive canal
Most posterior superior point of premaxilla
Staphilion
Maxillary tuberosity
Most superior midline point of pterygomaxillary fissure
Superior limit of foramen caecum
Sella turcica
Opisthocranion
Point of intersection between endocranial surface and root of orbit
Point of greatest curvature on superior endocranial surface
Opisthion
Basion
Root Block Landmarks
Mesial point of cervix of P_4
Mesial point of cervix of M_2
Distal point of cervix of M_3
Apex of mesial root of P_4
Apex of mesial root of M_2
Apex of distal root of M_3

of this covariation and the process by which this integrated variation is modulated during development remains unclear.

This paper forms part of an ongoing series of studies into the coordinated development and evolution of the masticatory system. The aim of this study is to test the hypothesis that variation in root length is correlated with facial length within *P. troglodytes*. The study assesses covariation between root length and the whole craniofacial skeleton. If the hypothesis is falsified, covariation between root length and alternative regions of the craniofacial skeleton will be explored.

Materials and Methods

The sample used in this study consists of lateral radiographs of male and female adult crania of *P. troglodytes* [9]. Crania were only selected that possess crowns without heavy wear, to avoid the effects of apical cementum deposition, and closed root apices, to ensure root growth was complete. The sex of the specimens was assessed based upon canine dimorphism observed on the lateral radiographs. The subspecific designation is not known for all specimens but it is clear that the sample contains more than one subspecies.

Twenty-three 2D landmarks (table 1) were digitised from each lateral radiograph using tpsDIG2 software [10]. Geometric morphometrics [11] were employed in this study to investigate morphological covariation. Generalised Procrustes Analysis (GPA) [12–15] was carried out using CoordGen6f software [16]. Subsequent to GPA, the landmarks were assigned to one of two blocks: one block of craniofacial landmarks, and one block of landmarks for the postcanine mandibular tooth roots. The analysis was restricted to the postcanine mandibular tooth roots because on lateral radiographs these roots are clearly visible and minimally distorted. The pattern of covariation between the landmarks in each of these blocks was assessed using partial least-squares analysis (PLS). PLS extracts vectors which maximize the covariance of shape variation within the blocks and between the blocks [14, 17]. Permutations (n = 999) were calculated to assess the statistical significance of the correlations between singular axes scores. The shape covariation on the vectors was visualised using Cartesian transformation grids calculated from thin plate splines [14]. All PLS analyses and visualizations were performed using tpsPLS software [18].

Results

The PLS analysis gives a strong and highly significant correlation coefficient for the first singular warps ($r = 0.76$; $p < 0.02$) which account for 64.7% of the total variation and show no relationship with sex. The correlation coefficients for all other singular warps are low and not statistically significant.

The shape variation associated with the statistically significant covariance identified on the first singular warp is shown in figure 2. A clear pattern of root length variation in the postcanine mandibular dentition can be seen in the root block on the first singular warp. At one extreme of the pattern, root length is close to equal in all tooth types examined, with a slight decrease in length in the roots of the more distal teeth. At the other extreme, there is a high degree of variation in root length along the tooth row, with the greatest length in the roots of the more mesial teeth, decreasing distally. The main shape variation in the craniofacial block along the first singular warp is in the relative height, particularly posteriorly, of the maxillary component of the facial skeleton. In addition there is correlated variation in the projection of the supraorbital region and the midface, nasoalveolar clivus of the premaxilla and cranial base angulation.

The covariation between the two blocks is such that a relative increase in posterior facial height, supraorbital projection and midfacial projection, a relative decrease in the length of the nasoalveoalr clivus, and a more flexed cranial base angle is correlated with variation in the relative length of the more mesial postcanine mandibular tooth roots.

Discussion

Previous comparative and experimental studies have demonstrated a correlation between tooth root length and facial length [6–8]. The present study used geometric morphometrics and PLS analysis to test this hypothesis on a sample of lateral cephalograms of *P. troglodytes*. The results of this study do not support the hypothesis that root length is correlated with facial length in this sample. The variation in root length along the tooth row is, however, correlated with facial height and a number of other craniofacial features.

By analyzing the roots of the tooth row of P_4-M_3 together in each specimen in this study, the findings demonstrate a previously unreported pattern of variation in root length along the postcanine tooth row. The observed variation ranged from a tooth row with relatively equal root lengths, to a tooth row with relatively long root lengths in the mesial teeth and root length decreasing distally to the shortest root length in the M_3.

The model supported by Riesenfeld and Siegel [6–8], whereby root length is related to facial protrusion, has no proposed mechanism to explain the developmental basis of this relationship. It was suggested that increased root length was found in teeth close to sites of growth such as sutures; however, such a model cannot apply directly to a structure without sutures, such as the mandible. So how is the covariance between

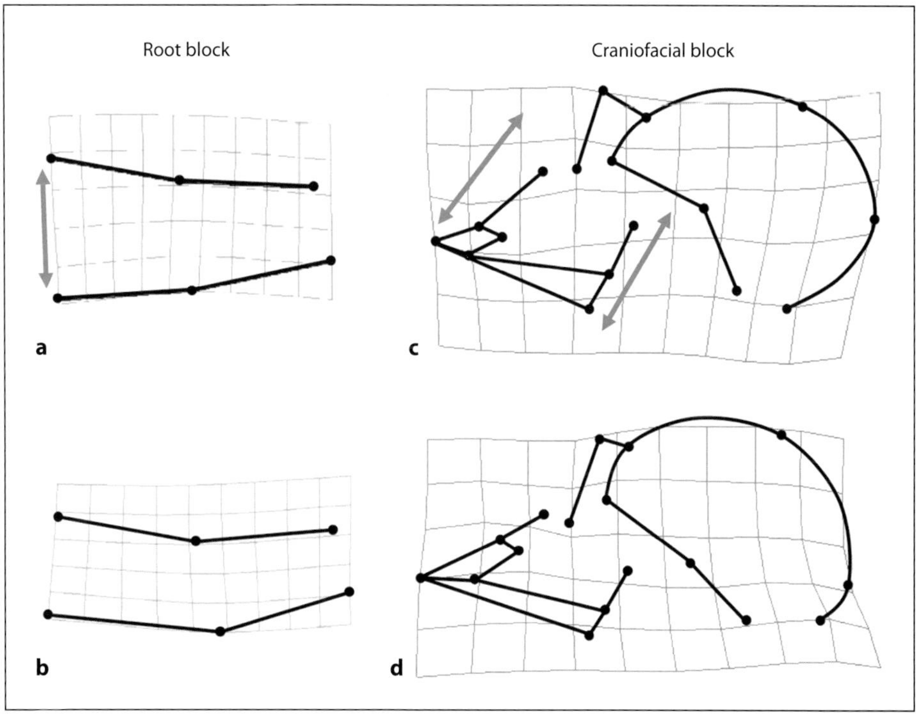

Fig. 2. Thin plate spline transformations of the shape covariation along the first singular warp. **a**, **c** The morphological covariation from one extreme of the vector and **b**, **d** from the other extreme. **a**, **b** The root block. **c**, **d** The craniofacial block.

mandibular root length and facial morphology observed in the present study modulated? Three hypotheses are discussed below:

(1) Given the strong relationship between tooth stability during food preparation and root length, the possibility that the differences in root lengths correlate with the forces that can be generated at each tooth should be considered. Muscle force production along the tooth row in humans, estimated using electromyographic data corresponds well to the mean pattern of root length variation along the tooth row [3] but there is no available data on variation in the pattern of force production along the tooth row that would allow this hypothesis to be tested in light of the patterns of variation in root length found in the present study. Spencer [3] proposed that root length and its associated stability, as sensed by the mechanoreceptors in the periodontal ligament, limit the magnitude of force that is generated at each tooth. This hypothesis does not provide an explanation of the covariation of root length in different tooth types and craniofacial morphology observed in this study.

(2) Rather than the muscle activity adapting to accommodate the ability of a root to resist the force generated at that position in the tooth row, the developing root of each tooth adapts plastically to the force it experiences in the period between coming into occlusion and closure of the root apex.

(3) A final hypothesis is that root length variation is a consequence of variation in the eruptive distance of each tooth. While the relationship

between root length and eruptive distance remains to be tested in a longitudinal cephalometric implant study, previous studies in *Macaca mulatta* have demonstrated that during development the inferior border of the mandibular corpus undergoes minimal remodeling whereas bone deposition occurs along the occlusal surface in a gradient that increases posteriorly. The overall effect of this remodeling pattern is a rotation of the mandibular occlusal surface [19, 20]. The alveolar remodeling rotation is intimated to be a compensatory process to maintain the position of the occlusal plane as the entire maxillary complex and mandible undergo a rotational displacement relative to the cranial base due to growth increases in the posterior facial height [20–22]. Therefore, if individuals undergo a relatively large increase in posterior facial height during growth, as demonstrated by the individuals at one extreme in the present study (fig. 2c), this model predicts that the compensatory remodeling of the mandibular alveolus would be greater than in an individual with a less pronounced ontogenetic increase in posterior facial height (fig. 2a, c). The degree of remodeling rotation on the occlusal surface of the mandible in *M. mulatta* decreases between M_1 and M_2 eruption [19] and so, due to the relative timing of molar development, the eruptive distance and the respective root length is predicted to be greatest in M_1 and decrease for each successive molar. This predicted pattern is indeed observed in this study in the individuals with a relatively greater posterior facial height (fig. 2a, c). The corollary that individuals with relatively short posterior facial heights will have little variation in root length along the tooth row is also observed (fig. 2b, d).

The findings of this study have demonstrated the pattern of variation in root length along the tooth row and how this covaries with posterior facial height. Three alternative, but not necessarily mutually exclusive, working hypotheses have been proposed that could account for the observed covariation but these remain to be tested in future studies.

References

1 Kovacs I: A systematic description of tooth roots; in Dahlberg AA (ed): Dental Morphology and Evolution. Chicago, University of Chicago Press, 1971, pp 211–256.
2 Spencer MA: Tooth root form and function in platyrrhine seed-eaters. Am J Phys Anthropol 2003;122:325–335.
3 Spencer MA: Force production in the primate masticatory system: electromyographic tests of biomechanical hypotheses. J Hum Evol1998;34:25–54.
4 Abbot SA: A comprehensive study of tooth root morphology in the great apes, modern man and early hominids; PhD thesis, University of London, 1984.
5 Kupczik KF: Tooth root morphology in primates and carnivores; PhD thesis, University of London, 2003.
6 Riesenfeld A: The effect of environmental factors on tooth development: an experimental investigation. Acta Anat 1970;77:188–215.
7 Seigel MI: The relationship between facial protrusion and root length in the dentition of baboons. Acta Anat 1972;83:17–29.
8 Riesenfeld A, Seigel MI: The relationship between facial proportions and root length in the dentition of dogs. Am J Phys Anthropol 1970;33:429–432.
9 Dean MC, Wood BA: A digital radiographic atlas of great apes skull and dentition; in Bondioli L, Macchiarelli R (eds): Digital Archives of Human Paleobiology. Milano, Consiglio Nazionale delle Ricerche, ADS Solutions, 2003, vol 3.
10 Rohlf FJ: tpsDIG2. Stony Brook, Department of Ecology and Evolution, State University, 1996.
11 Dryden I, Mardia KV: Statistical Shape Analysis. New York, Wiley, 1998.
12 Gower JC: Generalised procrustes analysis. Psychmetrika 1975;40:33–50.
13 Rohlf FJ, Slice D: Extensions of the procrustes method for the optimal superimposition of landmarks. Syst Zool 1990;39:40–59.
14 Bookstein FL: Morphometric Tools for Landmark Data. Cambridge, Cambridge University Press, 1991.
15 Goodall CR: Procrustes methods and the statistical analysis of shape. J R Stat Soc B 1991;53:285–340.
16 Sheets HD: IMP, Intergrated Morphometrics Package. Buffalo, Canisius College, 2001.
17 Rohlf FJ, Corti M: The use of two-block partial least-squares to study covariation in shape. Syst Zool 2000;49:740–753.

18 Rohlf FJ: tpsPLS Version 1.18. Stony Brook, Department of Ecology and Evolution, State University, 2001.
19 McNamara JA, Graber LW: Mandibular growth in the rhesus monkey (*Macacca mulatta*). Am J Phys Anthropol 1975;42:15–24.
20 Schneiderman ED: Facial Growth in the Rhesus Monkey. Princeton, Princeton University Press, 1992.
21 Enlow DH: Rotations of the mandible during growth; in McNamara JA (ed): Determinants of Mandibular Form and Growth. Ann Arbor, Center for Human Growth and Development, University of Michigan, 1975.
22 Enlow DH: Facial Growth, ed 3. Philidelphia, Saunders, 1990.

Dr. Samuel N. Cobb
Functional Morphology and Evolution Unit, Hull York Medical School, University of Hull
Hull, HU6 7RX (UK)
Tel. +44 0 1904 328873, Fax +44 0 1904 321696, E-Mail sam.cobb@hyms.ac.uk

Koppe T, Meyer G, Alt KW (eds): Comparative Dental Morphology.
Front Oral Biol. Basel, Karger, 2009, vol 13, pp 128–133

Root Growth during Molar Eruption in Extant Great Apes

Jay Kelley[a] · M. Christopher Dean[b] · Sasha Ross[c]

Departments of [a]Oral Biology and [c]Periodontics, College of Dentistry, University of Illinois at Chicago, Chicago, Ill., USA; [b]Department of Cell and Developmental Biology, University College London, London, UK

Abstract

While there is gradually accumulating knowledge about molar crown formation and the timing of molar eruption in extant great apes, very little is known about root formation during the eruption process. We measured mandibular first and second molar root lengths in extant great ape osteological specimens that died while either the first or second molars were in the process of erupting. For most specimens, teeth were removed so that root lengths could be measured directly. When this was not possible, roots were measured radiographically. We were particularly interested in the variation in the lengths of first molar roots near the point of gingival emergence, so specimens were divided into early, middle and late phases of eruption based on the number of cusps that showed protein staining, with one or two cusps stained equated with immediate post-gingival emergence. For first molars at this stage, *Gorilla* has the longest roots, followed by *Pongo* and *Pan*. Variation in first molar mesial root lengths at this stage in *Gorilla* and *Pan*, which comprise the largest samples, is relatively low and represents no more than a few months of growth in both taxa. Knowledge of root length at first molar emergence permits an assessment of the contribution of root growth toward differences between great apes and humans in the age at first molar emergence. Root growth makes up a greater percentage of the time between birth and first molar emergence in humans than it does in any of the great apes.

Dental development provides useful information for exploring a host of biological questions in humans, great apes and other primates. The timing of tooth eruption in particular is linked to matters as diverse as foraging ability and life history. While tooth eruption has received considerable attention, particularly in humans [1, 2], the progression of root development during the eruptive process is largely unexplored, particularly in apes. The first objective of this study, therefore, was to document the progression of root growth, and its variation, during molar eruption in great apes. However, there are reasons for this documentation beyond simply providing additional information on dental development in apes.

First molar (M1) eruption is particularly important because of the correlations between age at M1 gingival emergence and various life history attributes among living primates [3]. Therefore, determining age at M1 emergence is the best means of inferring the pace of life history among fossil primates. Because the M1 begins to form just prior to birth in higher primates [4, 5], for individuals that died during M1 eruption, age at M1 emergence equals age at death and can be determined by adding crown and root formation times calculated from the growth lines in the enamel and dentin of the M1 [6]. However, such individuals are rare, so there is a need to know how much root to include in these calculations using M1s that are past emergence and therefore have more fully formed roots. Thus, our second objective was to determine if the variation in the amount of root present at M1 emergence is sufficiently low that the range of root length to be used in the calculations, when added to crown formation times, would produce acceptably narrow ranges for the estimates of age at M1 emergence relative to the absolute mean ages of M1 emergence in great apes.

Knowing the progression of root formation also allows us to explore the developmental factors underlying the differences in age at M1 emergence among great apes and humans. This involves three components: crown formation time, root extension rates and the amount of root present at emergence. Crown formation times and root extension rates have been documented to varying degrees in great apes and humans [7–9]. The third component, the amount of root present at emergence and its variation, has not been systematically investigated for any species. In fact, age at M1 emergence is itself well documented only for humans [1, 2]. It is only very recently that we have begun to acquire reliable data for free-living great apes [6, 10, 11], so it is only now that we can begin to look at all the elements comprising the chronology of tooth eruption among apes and humans in a comparative fashion, and in the context of accurate M1 emergence ages. The final objective of this study, therefore, is to use the data on the amount of root present at emergence to evaluate the relative contributions of crown and root formation to differences in the age at M1 emergence among great apes and humans.

Materials and Methods

Great ape specimens with erupting first (M_1) or second (M_2) molars were examined in the osteology collections at several institutions (see 'Acknowledgments'). When possible, teeth were removed from the mandibles so that root lengths could be directly measured. Otherwise, root lengths were measured radiographically (periapical). The study was restricted to mandibular molars because of the difficulty of extracting maxillary molars for measurement, or of measuring root lengths on maxillary molars radiographically. In a number of specimens, the roots were measured both directly and radiographically to test the accuracy of the radiographic measurements. In fact, an additional aim of this study was to determine the accuracy of radiographic measurements in light of potential sources of error such as apical burnout and parallax, although the latter of these two sources of potential error was minimized by affixing the X-ray film directly to the lingual surface of the mandible. A paired t test with Bonferroni adjustment for 14 specimens in which roots were measured both directly and radiographically showed no significant difference. Therefore, direct and radiographic measurements were pooled, with preference given to the direct measurements for those specimens in which both types of measurements were made.

Three root length measurements were taken, mesial, distal and mesiobuccal, all recorded as the linear distance from the crown cervix to the apex of the developing root cone. Mesiobuccal root length was measured only on teeth that were removed from the mandible because of the difficulty in defining the mesiobuccal tooth cervix in radiographs. For measurements taken directly from the teeth, mesial and distal root lengths were measured at the buccolingual midpoint of the tooth. Measurements from x-rays were made directly from the film using digital calipers. For specimens that preserved left and right antimeres, both teeth were included in the data base. While data from antimeres cannot be considered completely independent, we observed that antimeres are sometimes at different stages of eruption. We assumed that root length is correlated with stage of eruption, or perhaps mandibular corpus depth, rather than being under some type of systemic control that is unrelated to these factors. For this reason, and to increase sample sizes, we treated antimeres as independent.

We limited the analysis to teeth in which the cusp tips were at or above the alveolar margin of the mandible but not fully rotated into the occlusal plane of the deciduous premolars, as this interval brackets the period of gingival emergence. Since it is not possible with osteological specimens to recognize teeth that were precisely at gingival emergence, teeth were divided into three stages, early, middle and late, based upon the presence and extent of protein stains on the cusps. In the absence of gingival tissue, the presence of protein stains is the only way to know that a tooth had emerged through the gingiva. Our observations revealed that such staining begins almost immediately upon gingival emergence, since stains

Table 1. Root lengths in mm at stage M (immediate postgingival emergence) in M_1 and M_2

	Gorilla			*Pan*			*Pongo*		
	mesial	distal	mesiobuccal	mesial	distal	mesiobuccal	mesial	distal	mesiobuccal
M_1									
n	12	13	8	8	8	7	4	4	4
Mean	6.86	4.73	5.11	4.83	3.83	3.76	5.15	3.88	3.62
Minimum	4.9	3.0	3.5	4.2	2.9	3.4	4.3	2.6	2.1
Maximum	7.9	7.2	6.7	5.3	4.9	4.2	6.1	5.9	5.4
SD	1.00	1.16	1.14	0.35	0.65	0.26	0.89	1.55	1.61
M_2									
n	4	4	4	7	6	4	4	3	3
Mean	7.38	4.28	4.52	7.04	6.13	6.29	6.58	4.40	4.35
Minimum	6.5	3.6	4.0	5.6	4.7	5.4	5.6	3.8	3.2
Maximum	8.0	5.0	5.5	8.2	7.6	7.1	8.5	5.6	6.3
SD	0.68	0.59	0.67	1.17	1.19	0.74	1.33	1.04	1.70

are sometimes present only at the very tips of cusps and demarcated by a clear gingival line. The eruption stages based on tooth staining were defined as follows:

Early (stage E)	0 cusps stained
Middle (stage M)	1–2 cusps stained
Late (stage L)	3–5 cusps stained

Thus, stage E includes teeth that were pre-emergent, stage M includes teeth that were recently post-emergent, and stage L includes teeth that were further past initial emergence but not yet fully in the occlusal plane. We defined stage M as the point of gingival emergence for determining root length at emergence. Since all teeth at this stage are slightly post-emergent, and since teeth classified as E that may have been on the verge of emergence are excluded, this scheme results in values for root lengths at emergence that are somewhat overestimated, but by an unknown amount. However, this error is likely to be relatively small given known root extension rates compared to the initial rate of retreat of the gingiva from the cusp tips after initial penetration.

Results

The progression of root development in the M_1 mesial and distal roots of *Gorilla, Pan* and *Pongo* is shown in figure 1a, b. The data are most complete for *Pan* and show that, for this genus, about 3 mm of root on average are added during eruption from the alveolar margin to near the occlusal plane.

M_1 and M_2 mesial, distal and mesiobuccal root lengths at stage M, which includes teeth just past initial gingival emergence, are shown in table 1 and figure 1c, d. M_1 Root lengths are greatest in *Gorilla* and fairly similar in *Pan* and *Pongo,* although sample sizes are quite small for *Pongo.* The results are less consistent for M_2, but sample sizes are uniformly small. In *Pan* and *Pongo,* M_2 roots are uniformly longer than those of M_1 at

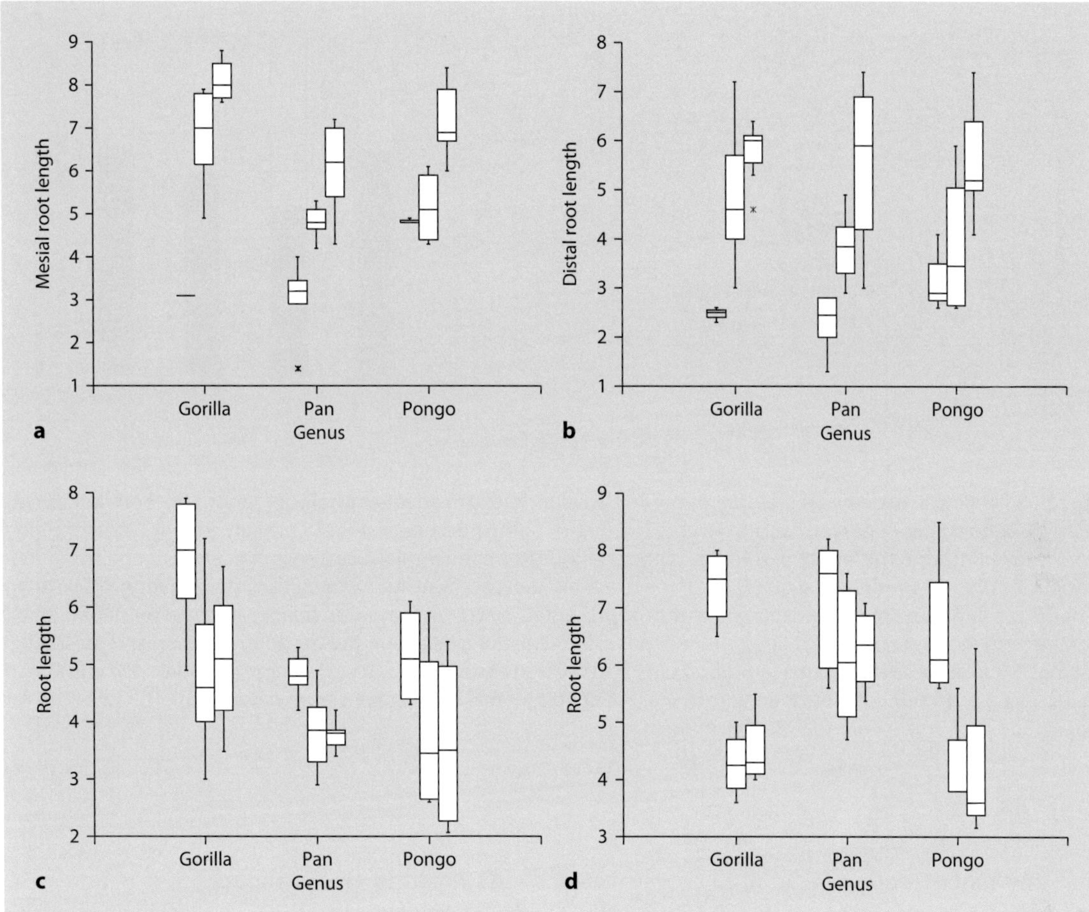

Fig. 1 a, b. Progression of root extension (mm) in the M_1 mesial (**a**) and distal (**b**) roots of extant great apes. For each taxon stage E is on the left, stage M is in the middle and stage L is on the right. **c, d** Root lengths at stage M (immediately after emergence) in M_1 (**c**) and M_2 (**d**). For each taxon the mesial root is on the left, the distal root is in the middle and the mesiobuccal root is on the right.

stage M. In *Gorilla* this is true for the mesial root but not for the distal or mesiobuccal roots, but, as before, sample sizes are small.

For both M_1 and M_2 in all species, the mesial root is longer than the distal or mesiobuccal roots at gingival emergence (stage M) (fig. 1c, d). Interestingly, in all species, the distal root and the mesiobuccal root are close to the same length at this stage. In fact, this relationship holds throughout the period of tooth eruption (fig. 2a) and is sufficiently strong (Pearson correlation coefficient = 0.913, $p < 0.01$) that distal root length can be used to estimate the length of the mesiobuccal root when the latter cannot be determined in radiographs.

Discussion

One of the questions posed in the Introduction was, is the variation in the amount of root present around gingival emergence sufficiently low

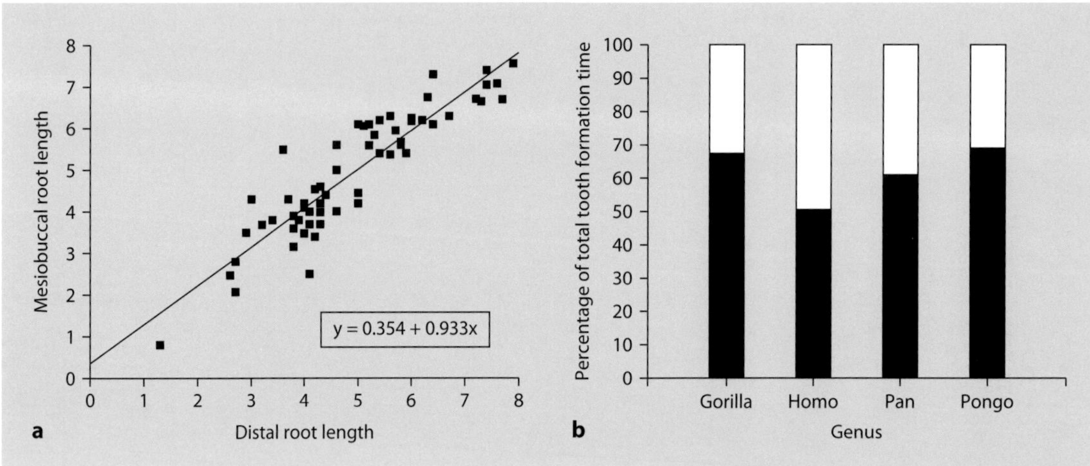

Fig. 2. **a** Linear least-squares regression of mesiobuccal root length against distal root length for both M_1 and M_2 at all eruption stages in *Gorilla, Pan* and *Pongo.* The length of the mesiobuccal root normally cannot be determined from X-rays. Knowing the length of this root is important because the mesiobuccal cusp is the first to begin mineralization, making the mesiobuccal quadrant of the tooth the most accurate for histological determination of age at M_1 emergence. **b** Approximate percentages of time represented by crown formation (black) and root formation (gray) at M_1 emergence in extant great apes and humans. All calculations based on the mesiobuccal cusp and mesiobuccal root. Sources for average cusp formation time and root extension rates: *Pan* [7], *Homo* [8], and *Gorilla* and *Pongo* [9, 11]. Average amount of mesiobuccal root present at emergence in great apes from data in table 1; same data for *Homo* [8].

that any molar tooth can be used to estimate age at M1 emergence histologically by adding crown formation time to the time to form the amount of root that would have been present at emergence? The ranges defined by the lower and upper 95% confidence limits for the length of the mesial root at stage M in *Pan* and *Gorilla* are, respectively, 0.57 and 1.28 mm. When divided by the average root extension rates over the approximate interval for stage M (*Pan*: 4th to 5th mm of root, 8.9 μm/day [7]; *Gorilla*: 6th to 8th mm or root, 13.0 μm/day [9, 11]) these ranges equal just over two and three months of root formation, respectively. These time brackets would produce reasonably narrow ranges of the estimated age at emergence relative to the likely average age of M_1 emergence of between 3.5 and 4 years of age in wild chimpanzees and gorillas [10, 11].

As noted in the Introduction, it is now possible to present a preliminary assessment of the factors underlying the differences in age at M1 emergence in great apes (*Pan* and *Gorilla*: ~3.5–4.0 years; *Pongo*: ~4.5–5.0 years; humans: ~5.6–6.0 years). Figure 2b demonstrates that the delay in humans relative to the great apes is due at least in part to a relatively longer period of root formation, resulting from both a lower average root extension rate (~7 μm/day) as well as greater root length at emergence in humans (~8 mm) [9] compared to great apes, as demonstrated by the results presented here.

Acknowledgments

We thank the following institutions for access to specimens: National Museum of Natural History, Washington,

D.C., Cleveland Museum of Natural History, Field Museum of Natural History, Chicago, Ill., American Museum of Natural History, Powell-Cotton Museum, Museum für Naturkunde der Humboldt-Universität zu Berlin, Bayerische Staatssammlung für Zoologie, Munich, Natural History Museum, London, Royal College of Surgeons, London. Aspects of this study were supported by a grant to MCD from the European Commission's Research Infrastructure Action via the SYNTHESYS Project.

References

1 Smith RJ, Gannon PJ, Smith BH: Ontogeny of australopithecines and early *Homo*: evidence from cranial capacity and dental eruption. J Hum Evol 1995;2:155–168.
2 Liversidge HM: Worldwide variation in human dental development; in Thompson JL, Nelson A, Krovitz G (eds): Growth and Development in the Genus *Homo*. Cambridge, Cambridge University Press, 2003, pp 73–113.
3 Smith BH: Dental development as a measure of life history in primates. Evolution 1989;43:683–688.
4 Dean MC: The developing dentition and tooth structure in hominoids. Folia Primatol 1989;53:160–177.
5 Beynon AD, Dean MC, Reid DJ: Histological study on the chronology of the developing dentition in gorilla and orangutan. Am J Phys Anthropol 1991;86:189–204.
6 Kelley J, Schwartz GT: Histologically determined age at first molar emergence in *Pongo pygmaeus*. Am J Phys Anthropol Suppl 2005;S40:128.
7 Smith TM, Reid DJ, Dean MC, Olejniczak AJ, Martin LB: Molar development in common chimpanzees (*Pan troglodytes*). J Hum Evol 2007;52:201–216.
8 Mahoney P: Intraspecific variation in M_1 enamel development in modern humans: implications for human evolution. J Hum Evol 2008;55:131–147.
9 Dean MC, Vesey P: Preliminary observations on increasing root length during the eruptive phase of tooth development in modern humans. J Hum Evol 2008;54:258–271.
10 Zihlman A, Bolter D, Boesch C: Wild chimpanzee dentition and its implications for assessing life history in immature hominin fossils. Proc Nat Acad Sci USA 2004;101:10541–10543.
11 Kelley J, Schwartz GT: New ages at first molar emergence in extant great apes and a reassessment of early hominin first molar emergence ages. Am J Phys Anthropol Suppl 2009;S48:164.

Jay Kelley, PhD
Department of Oral Biology (m/c 690), College of Dentistry, University of Illinois at Chicago
801 S. Paulina
Chicago, IL 60091 (USA)
Tel. +1 312 996 6054, Fax +1 312 996 6044, E-Mail jkelley@uic.edu

Koppe T, Meyer G, Alt KW (eds): Comparative Dental Morphology.
Front Oral Biol. Basel, Karger, 2009, vol 13, pp 134–135

Clinical Aspects of Dental Morphology: An Introduction

Alan Brook

University of Liverpool, School of Dental Sciences, Liverpool, UK

In this section of this volume, we have a series of refereed research papers which relate both to clinical setting and to dental morphology. These papers not only reflect existing knowledge but also point to exciting future developments.

First, Townsend's group examines how studies of twins can inform our understanding of dental morphology. Two metaphors are considered. The first views teeth as being twins related to antimeres and isomers. The other metaphor represents decision making by cells during dental development as 'an epigenetic landscape', in which the decision making is like a ball rolling in an undulating landscape of interconnecting hills and valleys. In this way, dental phenotypic differences may arise within and between monozygotic co-twins from relatively minor temporo-spatial effects during development.

Differences in dental crown size between antimeric pairs of monozygotic co-twins have been demonstrated using manual methods and 2D and 3D imaging systems. Future studies will exploit the new imaging techniques, providing further detail of tooth morphology, and multivariate analysis to examine the causes of the similarities and differences within and between the dentitions of twins using the two metaphors. Epigenetic, as well as genetic and environmental, sources of variation will be explored.

Second, Smith and coworkers look at applying a novel superimposition technique to tooth morphology data derived from 3D measurement of the dental study models of monozygotic co-twins. The 3D measurement is providing new variables and shape information and differences between twins can be highlighted by accurate superimposition of corresponding teeth. This study aims to develop a 'non-best fit' (Synetic) method of superimposition to identify differences in the teeth of co-twins. The initial results suggest that the new method was more reliable than a conventional best fit approach for superimposing tooth crowns.

Third, Horrocks and coworkers report a 3D comparison of the modifying effects of the familial genetic contribution in Turner (XO) syndrome. Comparison of individuals affected by chromosomal anomalies with controls has shown the differential action of X and Y chromosomes during dental development. Individuals with Turner syndrome have smaller teeth than controls. This pilot study used a 3D system to measure the dentition of 10 individuals with Turner syndrome and 10 unrelated female controls. The reliability of the method was high and it was effective in measuring differences between the two groups quantitatively. The main study involving 35 individuals with Turner syndrome, 35 first-degree unaffected female relatives and 35 female controls is being completed. The aim is to measure surface morphology, volume and contours of the dentition to quantify the influence of Turner syndrome on dental development and to assess any modifying effects of familial genetic contribution.

Fourth, Liversidge addresses a problem which has important clinical, sociological and public health implications, that of age estimation based on permanent tooth formation. The aims of this study were to describe maturity data of permanent tooth formation from a large sample of dental patients in London and to adapt these for estimating age, comparing mean age dental formation stages between ethnic groups and sexes. Over 1,000 panoramic radiographs of similar numbers of healthy White and Bangladeshi dental patients aged between 2 and 22 years were scored. The results demonstrated an accurate method of estimating age from developing permanent mandibular teeth.

These papers demonstrate the value and application of important new techniques and methodologies to important clinical and dental morphology research questions.

Prof. Alan Brook
School of Dental Sciences
University of Liverpool
Edwards Building, Daulby Street
Liverpool L69 3GN (UK)
Tel. +44 151 706 5665, Fax +44 151 706 5809, E-Mail A.H.Brook@liv.ac.uk

Koppe T, Meyer G, Alt KW (eds): Comparative Dental Morphology.
Front Oral Biol. Basel, Karger, 2009, vol 13, pp 136–141

How Studies of Twins Can Inform Our Understanding of Dental Morphology

Grant Townsend[a,b] · Toby Hughes[a] · Michelle Bockmann[a] · Richard Smith[b] · Alan Brook[b]

[a]School of Dentistry, University of Adelaide, Adelaide, S.A., Australia; [b]School of Dental Sciences, University of Liverpool, Liverpool, UK

Abstract

Two metaphors are presented to highlight concepts that could lead to a paradigm shift in dental studies of twins. The first, derived from the Song of Solomon in the Bible, refers to teeth as being twins. This viewpoint emphasises that each tooth should be viewed as a paired structure, not only with its antimere (within the same arch) but also with its isomer (in the opposing arch). The other metaphor provided by Waddington in 1957 is visual and involves 'an epigenetic landscape' that represents the processes of decision-making by cells during development. It likens the different stages of cellular decision-making to a ball rolling down an undulating landscape of interconnecting hills and valleys. This viewpoint helps to explain how distinct differences in dental phenotypes may arise both within and between monozygotic (MZ) co-twins due to relatively minor temporospatial effects during development. Measurements of maximum mesiodistal diameters of teeth in a pair of MZ twins, using calipers and also 2D and 3D imaging systems, have demonstrated that differences in dental crown size occur between antimeric pairs and between corresponding teeth of MZ co-twins. By defining new dental phenotypes that provide more comprehensive descriptions of tooth size and shape, and by drawing on the metaphors described, we are confident of providing new insights into the reasons for observed similarities and differences within, and between, the dentitions of twins. Our approaches will focus on multivariate analyses that take into account the paired arrangement of teeth and also explore epigenetic, as well as genetic and environmental, sources of variation.

In preparing this paper, a Google search was performed using the key words 'teeth and twins'. A reference was found to the Song of Solomon or Song of Songs, one of the books in the Hebrew Bible. The fourth chapter of this song consists of a series of metaphors, including the following verse:

'Thy teeth are like a flock of sheep that are even shorn, which came up from the washing; whereof every one bears twins, and none is barren among them.'

We believe that the metaphorical nature of this reference to teeth as twins is particularly relevant. When considered together with the importance of metaphors in shaping scientific thought and discovery, including so-called 'paradigm shifts' [2, 3], the phrase opens up a new

All authors are associated with the International Collaborating Centre in Oro-facial Genetics and Development.

perspective into how future studies of twins can inform our understanding of dental morphology. Rather than using the phrase 'teeth and twins', we suggest that our focus should be on the metaphor 'teeth are twins'.

This viewpoint emphasises that each tooth should be viewed as a paired structure, not only with its antimere (within the same arch) but also with its isomer (in the opposing arch). We propose that future studies of dental morphology, based on data derived from twins, are likely to be most informative if they adopt a multivariate approach that explores how genetic, epigenetic and environmental factors influence both antimeric and isomeric covariation within the human dentition. The metaphor also reminds us that the dentitions of twins provide a unique model system to explore the early determination of body symmetry in humans, including the phenomenon of mirror-imaging.

We also wish to refer to a marvellous visual metaphor provided by Waddington [1] that helps to explain how distinct differences in dental phenotypes may arise within and between monozygotic (MZ) co-twins. Waddington proposed the concept of 'an epigenetic landscape' to represent the processes of decision-making by cells during development. He likens the different stages of cellular decision-making during development to a ball rolling down an undulating landscape of interconnecting hills and valleys. At various stages on this 'landscape', the rolling ball (or cell) can take specific permitted trajectories leading to different outcomes or cellular fates. This metaphor seems remarkably appropriate to us when considering the developmental processes involved in odontogenesis, with minor temporo-spatial variations in cellular interactions apparently having the potential to lead to quite different phenotypic outcomes. Thus, minor differences in the 'epigenetic landscape' between right and left sides may lead to distinct phenotypic differences between antimeric teeth. Furthermore, differences in the 'epigenetic landscape' between MZ co-twins may also lead to differences in their dental phenotypes.

The Classical Twin Model and Its Assumptions

MZ co-twins share the same genes whereas dizygotic (DZ) co-twins share only half of their genes on average. By assuming that both types of twins are sampled from the same gene pool and that similar environmental factors affect them, one can estimate the relative contributions of genetic and environmental influences to observed variation in different traits. However, there are several assumptions underlying the classical twin approach and these have not been tested fully in many studies. For example, the mean values for the trait under investigation should not differ between zygosity groups. Total variance within zygosities should also be equal for the model to hold, as heterogeneity of total variance suggests that environmental factors are not equal for MZ and DZ twins. Environmental covariances should also be equal, with heritability estimates being inflated if environmental covariance is greater in MZ twins than DZ twins. All of these assumptions should be tested statistically prior to calculating genetic and environmental contributions to phenotypic variance.

Other Twin Models

There are several other research methods, apart from the classical twin model, that can be used when studying twins [4]. One approach that overcomes the problem of possible confounding effects due to common family environment is to study MZ twins who were separated shortly after birth and reared in separate homes. Assuming that their adoption placement is not influenced by trait-related environmental factors, the similarity of reared-apart MZ co-twins can be attributed to shared genes alone [5].

Another design is the MZ co-twin model which involves comparison of MZ twin pairs, where co-twins have been exposed to different environmental effects. The MZ co-twin model can also be used to make inferences about the relative contributions of genetic and environmental factors to phenotypes for which the twins are discordant. Differences in the number and position of missing teeth between MZ co-twins who have been confirmed to be monozygotic based on DNA analyses, raise the likelihood that environmental and/or epigenetic influences during development can lead to quite distinct differences in dental development [6, 7].

Special Features of the Twinning Process

One of the main assumptions of the classical twin model is that the findings from these studies can be extrapolated to the general population. However, there are some who question whether this is appropriate, even for dental variables [8, 9]. The twinning process is a special event, leading to the concurrent development of two individuals in utero. There are also special circumstances surrounding the birth of twins and their peri-natal development. Although some claim that the potentially harmful effects of twin gestation have been exaggerated [10], a large percentage of twins may not develop past 16 weeks post-conception, leading some researchers to refer to a so-called 'vanishing twin syndrome' [11]. Another special phenomenon that can occur in MZ twin pairs is 'mirror imaging', where one twin 'mirrors' the other for one or more features. Given that there is some preliminary evidence that mirror-imaging may be related to the timing of the twinning process and therefore the type of placentation [12], information on chorion type of MZ twins would be extremely valuable in any future study of mirror-imaging in twins.

Application of Imaging Approaches to Compare Tooth Size in a Pair of MZ Twins

We have shown previously that there can be large differences in dental phenotypic expression between MZ twin pairs [6, 7]. For example, the maximum mesiodistal crown diameters of permanent maxillary central incisor teeth were shown to differ by 0.5 mm or more in 5% of MZ twin pairs. This magnitude of difference is greater than can be attributed to measurement error when using hand callipers, indicating that 'real' differences in tooth size are discernible between individuals with the same genotype.

We present data for maximum mesiodistal crown diameters of the upper central incisors derived from one pair of monozygotic twins (T50 A and B). Measurements were obtained directly from dental casts using digital callipers with sharpened beaks and also by using the 2D and 3D imaging systems developed by Brook, Smith and colleagues at the University of Liverpool [13, 14].

Figure 1 shows labial views of maxillary dental models for the MZ co-twins. Maximum mesiodistal crown diameters for T50A, based on several separate recordings using callipers, were 8.0 mm for the upper right central incisor and 7.8 mm for the upper left central incisor. Corresponding measurements for the T50A based on scanned 2D images were 8.1 and 7.6 mm. For T50B, the calliper measurements were 8.5 mm for the upper right central incisor and 8.6 mm for the upper left central incisor. Corresponding measurements from 2D images were 8.5 mm for the right side and 8.9 mm for the left. When 3D images were used to record the projected maximum mesiodistal crown diameters of the central incisors in both co-twins, the values obtained were as follows: for T50A, 8.0 mm for the upper right and 7.6 mm for the upper left; and for T50B, 8.6 mm for the upper right and 8.6 mm for the upper left.

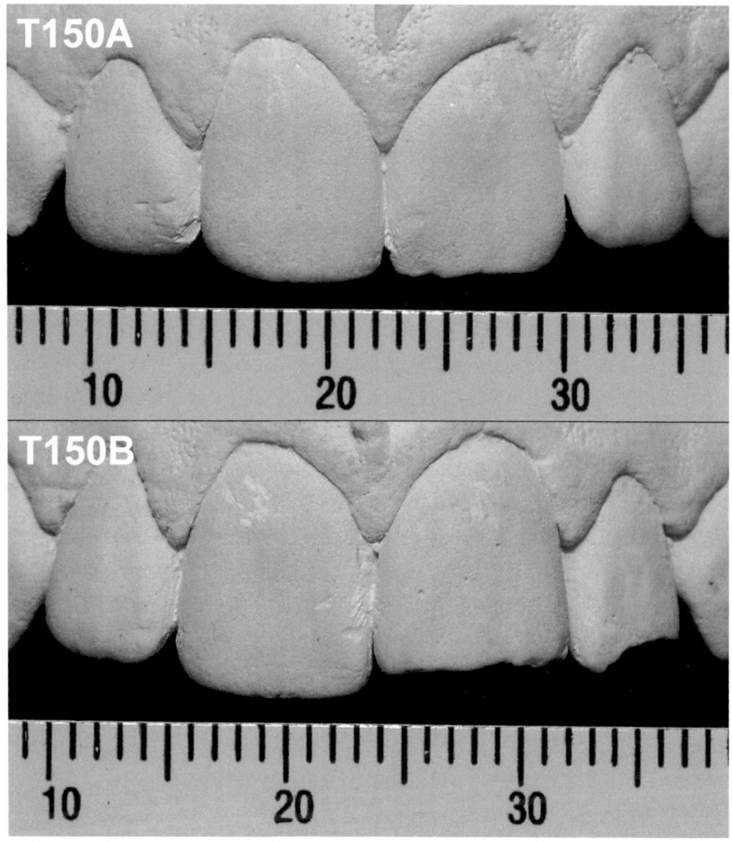

Fig. 1. Labial view of the maxillary dental models of a pair of monozygotic twins. The right and left central incisors were measured with hand-held callipers, and also using 2D and 3D imaging systems. Differences in the size of these antimeric teeth were noted within and between the co-twins that were larger than discrepancies expected due to measurement error.

Interpreting Our Findings and Linking the Two Metaphors

By taking advantage of the paired arrangement of teeth within the dental arches and also by applying different twin models, such as the MZ co-twin design, we believe that it will be possible to gain new insights into the roles of genetic, epigenetic and environmental influences on human dental development. Furthermore, it is now possible to define new dental phenotypes that can be measured accurately and reliably. These new phenotypes, including perimeters, areas and volumes, have the potential to provide better insights into the factors contributing to variation in dental morphology than can be gained by using traditional crown diameters alone.

The example that we have provided of one pair of MZ twins highlights the fact that there can be quite distinct differences in the size of corresponding teeth between co-twins, even though they share the same genes. There is also evidence of asymmetry between antimeric teeth within individuals. Given the accuracy and reliability of the measuring systems used, differences of the order of 0.5 mm are unlikely to be due to measurement errors.

Although molecular geneticists have tended to focus on methylation and acetylation of DNA when referring to epigenetics, there is growing recognition of the need for a broader interpretation of the term [15, 16]. We use the term 'epigenetics' in the broad sense proposed originally by Waddington [1] to refer to processes by which genotype gives rise to phenotype. This broader view enables us to describe the interactive processes that occur between cells at the local tissue level during dental development as epigenetic events, in addition to those that may operate directly on DNA.

Conclusion

Although it has been established that there is a strong genetic contribution to variation in dental crown size and shape, we propose that epigenetic and environmental factors can also contribute to observed variation. We are exploring the roles of these factors by using 2D and 3D imaging approaches in large samples of twins of known zygosity. In doing so, we are defining new dental phenotypes that reflect underlying odontogenic processes and we will maintain a focus on teeth as paired structures. Recalling the metaphors raised in this paper, we will be constantly reminding ourselves that 'teeth *are* twins' – a perspective that will hopefully lead to new insights into the how epigenetic influences, as well as genetic and environmental factors, lead to variations in human dental morphology.

Acknowledgements

We acknowledge the support of the following funding agencies: National Health and Medical Research Council of Australia; Australian Dental Research Foundation; Wellcome Trust, UK. We also wish to express our sincere thanks to all of the twins and their families who have participated in our studies and to Sandy Pinkerton who managed the twin studies in Adelaide from their commencement in the 1980s until her retirement in 2008.

References

1 Waddington CH: The Strategy of Genes: A Discussion of Some Aspects of Theoretical Biology. London, Allen & Unwin, 1957.
2 Kuhn TS: The Structure of Scientific Revolutions. Chicago, Chicago University Press, 1962.
3 Ortony A: The Shift from Metaphor to Analogy in Western Science, ed 2. Cambridge, Cambridge University Press, 1994.
4 MacGregor AJ, Snieder H, Schork NJ, Spector TD: Twins: novel uses to study complex traits and genetic diseases. Trends Genet 2000;16: 131–134.
5 Bouchard TJ Jr: Twins reared apart and together: what they tell us about human diversity; in Fox SW (ed): The Chemical and Biological Basis of Individuality. New York, Plenum, 1984, pp 147–178.
6 Townsend G, Richards L, Hughes T, Pinkerton S, Schwerdt W: Epigenetic influences may explain dental differences in monozygotic twin pairs. Aust Dent J 2005;50:95–100.
7 Townsend G, Hughes T, Richards L: The dentitions of monozygotic twin pairs: focusing on the differences rather than the similarities; in Zadzinska E (ed): Current Trends in Dental Morphology Research. Refereed Proceedings, 13th International Symposium on Dental Morphology. Lodz, University of Lodz Press, 2005, pp 337–352.
8 Boklage CE: Developmental differences between singletons and twins in distributions of dental diameter asymmetries. Am J Phys Anthropol 1987;74:319–331.
9 Harris EF: Genetic and environmental influences on tooth crown diameters in twins. 83rd General Session, International Association for Dental Research, Baltimore, 2005.
10 Martin N, Boomsma D, Machin G: A twin-pronged attack on complex traits. Nat Genet 1997;17:387–392.
11 Landy HJ, Keith LG, Keith DM: The vanishing twin. Acta Genet Med Gemellol (Roma) 1982;31:179–194.
12 Townsend G, Richards LC, Brown T: Mirror imaging in the dentitions of twins – what is the biological basis?; in Brown T, Molnar S (eds): Craniofacial Variation in Pacific Populations. Adelaide, The University of Adelaide, Anthropology and Genetics Laboratory, 1992, pp 67–78.
13 Brook AH, Smith RN, Elcock C, Al-Sharood MH, Shah AA, Khalaf K, Robinson DL, Lath DL, Karmo M: The measurement of tooth morphology: development and validation of a new image analysis system; in Zadzinska E (ed): Current Trends in Dental Morphology Research. Lodz, University of Lodz Press, 2005, pp 475–482.

14 Smith R, Zaitoun H, Coxon T, Karma M, Kaur G, Townsend G, Harris E, Brook A: Defining new dental phenotypes using 3-D image analysis to enhance discrimination and insights into biological processes. Archs Oral Biol 2008;in press.

15 Van de Vijver G, Van Speybroeck L, de Waele D: Epigenetics: a challenge for genetics, evolution, and development? Ann NY Acad Sci 2002;981:1–6.

16 Goldberg AD, Allis CD, Bernstein E: Epigenetics: a landscape takes shape. Cell 2007;128:635–638.

Prof. Grant Townsend
School of Dentistry, The University of Adelaide
Adelaide SA 5005 (Australia)
Tel. +61 883035968, Fax +61 883033444, E-Mail grant.townsend@adelaide.edu.au

Koppe T, Meyer G, Alt KW (eds): Comparative Dental Morphology.
Front Oral Biol. Basel, Karger, 2009, vol 13, pp 142–147

Synetic Superimposition of Dental 3D Data: Application in Twin Studies

Richard N. Smith[a] · Grant Townsend[a,c] · Ke Chen[b] · Alan Brook[a]

[a]School of Dental Sciences and [b]Division of Applied Mathematics, University of Liverpool, Liverpool, UK; [c]School of Dentistry, University of Adelaide, Adelaide, South Australia

Abstract

Background: Three-dimensional measurement of dental morphology is providing new variables and shape information not available previously, and the reliability of these data has proved to be substantial. Accurate superimposition of 2D and 3D data-sets has several applications in dental research when making comparisons of similar structures. For example, two data-sets of the same object can be superimposed to highlight differences apparent in limited region(s) of the tooth crown. When significant regions represented by the two data-sets are identical, registration of the images can be achieved by standard superimposition methods. However, for comparisons of crown morphologies between monozygotic and dizygotic co-twins, the data sets are similar but not identical so a new approach to superimposition has been developed. **Aim:** To develop a non 'best fit' (Synetic) method for the 3D superimposition of non-similar objects that forms one comparable interface providing enhanced methodology for the analysis of differences within the dentition of twins. **Method:** A minimised least-squares registration approach is followed by diffusion based registration to provide global minimisation between points that is not based on a best fit algorithm. This process is linear and therefore the method ensures uniqueness of the superimposition. **Results:** Initial results indicate a reliable method producing only one output as opposed to best fit approaches that may generate a different output each time. **Conclusion:** Diffusion-based registration offers a more reliable approach to superimposing non-identical objects, such as the tooth crowns of monozygotic co-twins, than conventional best-fit methods.

There are two main classifications for twins, these being monozygotic (MZ) and dizygotic (DZ). MZ twins have developed when a single egg is fertilized to form one zygote (MZ) which then divides into two separate embryos. They have in theory identical genotypes. DZ twins usually occur when two fertilized eggs are implanted in the uterine wall at the same time and share, on average, half their genes. Complex developmental interactions controlled by multifactorial inheritance and influenced by epigenetic and environmental factors are responsible for the final morphology of dental features [1–3]. The development of accurate and reliable mathematically based methods of comparing tooth surfaces should help unravel the contribution of genetic and environmental factors to phenotypic variation within the dentition.

The study of twins is a particularly valuable approach to quantify the degree of involvement of each of the contributing factors to observed variation, providing an opportunity to specifically explore epigenetic and environmental influences [4]. This study demonstrates the advantages of the dentition as a model for reflecting general developmental processes. Detailed description and measurement of specific dental landmarks, such as cusp positions, tooth size and shape, in 3D, can provide important new insights into the complex process of odontogenesis.

Comparing the similarity of dental morphology between twin pairs is of great research interest but it is also very demanding. To superimpose the dentitions and study the differences between MZ twin pairs in 3D would provide many new indicators and measures of dental variation, as they share 100% of their genetic material. Epigenetic factors and environmental differences between MZ co-twins result in morphological differences, but locating where the differences occur and quantifying the differences is particularly challenging.

The usual approach to 3D superimposition assumes that the subject matter (e.g. a tooth) is the same item that has undergone a process (e.g. erosion or restoration) and has changed part of the surface only but, in the main, the tooth remains the same. In this case, it is reasonable to use a 'best fit' mathematical approach which is sufficient to produce the desired comparison. When making comparisons between twins, 'best fit' can produce reasonable results in some cases [5], but the approach is dependent upon ambiguous mathematics that does not provide uniqueness in the final superimposed objects. This may give weighting towards more similar areas to the detriment of other regions.

In this paper, we outline an adaptation of the image registration software to facilitate non-identical object twin comparisons. This novel application has been named Synetic superimposition, derived from the Greek word 'synectikos' which means 'bringing forth together' or ' bringing different things into unified connection'.

Method

The procedure implements image diffusion based registration techniques and then demonstrates how to adapt the surface data, stored in the measurement software's Cloud STL format, to a form suitable for using the imaging technique. Then, the registered result is converted back to Cloud's STL format for further displays and manipulations (Cloud Software, Robin Richards, MedPhys, London, UK).

Ideas and Formulation from Image Registration

Image registration is the process of spatially aligning two or more images of the same object obtained at different times or from different viewpoints or by different imaging machineries as in multi-modality imaging [6–8].

For continuous variables, be represented by compactly supported functions $R,T\colon \Omega \subset R^2 \to V \subset R_0^+$. For each pixel $\mathbf{x} = (x_1, x_2)^*$, denote by $\varphi = \varphi(\mathbf{x})\colon \Omega \to \Omega$ the unknown coordinate transformation that produces the alignment between the reference R and the transformed version of the template $F = T(\varphi(\mathbf{x}))$. We hope to achieve $F \approx R$ by minimizing the least-squares functional (with respect to φ):

$$D[\boldsymbol{\varphi}] = \frac{1}{2}\int_\Omega \left(T(\boldsymbol{\varphi}(\mathbf{x})) - R(\mathbf{x})\right)^2 d\mathbf{x}.$$

Unfortunately, directly minimising this functional is an inverse problem which does not have a unique solution, even when φ is an affine (linear) transform [7].

One popular solution for solving an inverse problem is to add a regularization term. Using the so-called L_2 norm of the gradient of $u(\mathbf{x}) = \mathbf{x} - \varphi(\mathbf{x})$ the regularized functional is the following:

$$J[u] = \frac{1}{2}\int_\Omega \left(T(\mathbf{x} - u(\mathbf{x})) - R(\mathbf{x})\right)^2 d\mathbf{x} + \frac{\alpha}{2}\int_\Omega \left(|\nabla u_1|^2 + |\nabla u_2|^2\right) d\mathbf{x},$$

where α is a positive parameter which may be chosen by the L-curve method. Here once the transformation $u(\mathbf{x}) = (u_1(\mathbf{x}), u_2(\mathbf{x}))$ is found, the registered image will be $T(\mathbf{x}-u(\mathbf{x}))$. In practice, when R and T are given as discrete matrices (images), interpolation must be used to work out $T(\mathbf{x}-u(\mathbf{x}))$ whenever $u(\mathbf{x})$ is not an integer pair (vector). We remark that solving the above functional $J[u]$ is equivalent to solving the Euler-Lagrange partial differential equations:

$$\Delta \mathbf{u}(\mathbf{x}) + \alpha\big(\mathbf{T}(\mathbf{x} - \mathbf{u}(\mathbf{x})) - R(x)\big) \cdot \nabla \mathbf{T}(\mathbf{x} - \mathbf{u}(\mathbf{x})) = 0,\ \mathbf{x} \in \Omega,$$

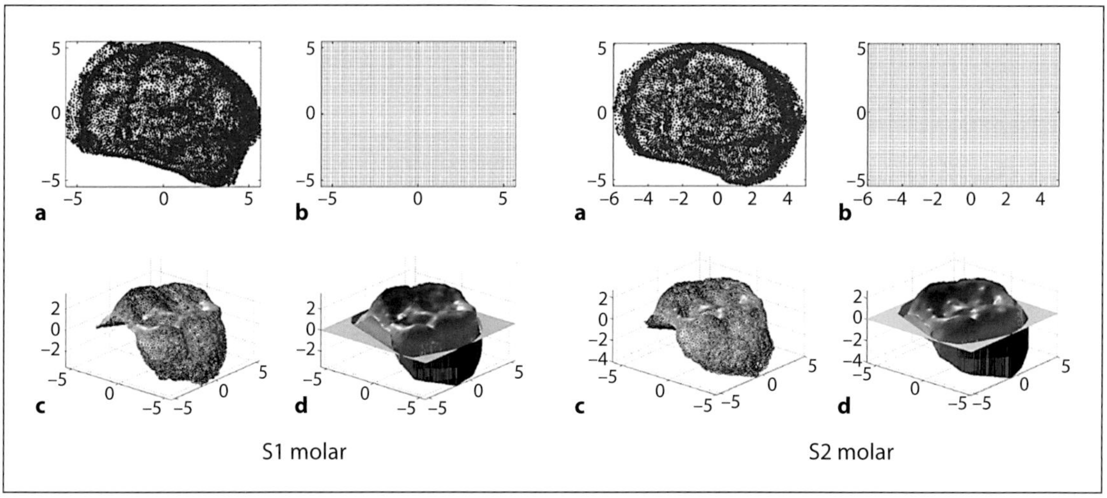

Fig. 1. Result for the upper 1st molar of one MZ co-twin, S1, and for the other co-twin, S2.

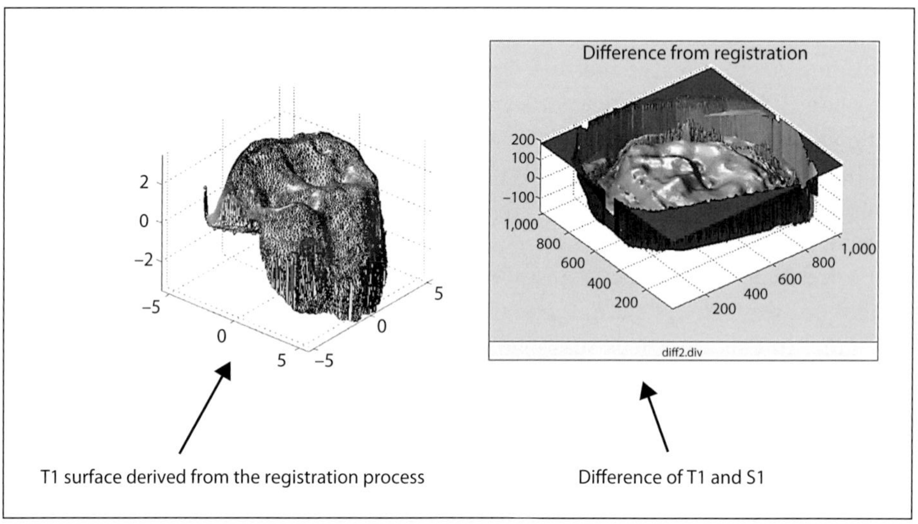

Fig. 2. The T1 surface derived from the registration process along with the difference of T1 and S1.

which resemble the diffusion equations from fluid simulation (hence the name diffusion registration). Full computational details are available [7, 8].

Conversion of Surface Data into Image-Like Data
A pair of surfaces S1 and S2 from the molar crowns of a pair of MZ twins which are stored in STL format used by Cloud software. The following simple procedure has been employed for the data conversion:

(1) Extract the triple (x_i, y_i, z_i), via GMESH format, of each of the two surfaces and their connectivity table E which are combined to form the 3D surfaces of teeth.

(2) Work out the smallest rectangular box Ω that contains all (x_i, y_i) of both surfaces.

(3) Divide Ω into a fine resolution and uniform distribution of pixels like coordinate points (X_j, Y_j).

(4) Interpolate (x_i, y_i, z_i) onto the new pixel positions to obtain all (X_i, Y_i, Z_i) which may be treated as an image

Table 1. Intra-operator and inter-operator reliability, based on Intra-Class Correlation Coefficient values, using 3-D image analysis

3-D image analysis	(Upper left central) variables						
	MD	LL	IG	MDa	LLa	IGa	SA
Intra-reliability 3D	0.887	0.907	0.985	0.921	0.95	0.976	0.996
Inter-reliability 3D	0.918	0.561	0.975	0.759	0.704	0.967	0.988
	(Upper right lateral) variables						
	MD	LL	IG	MDa	LLa	IGa	SA
Intra-reliability 3D	0.982	0.925	0.974	0.90	0.883	0.958	0.991
Inter-reliability 3D	0.940	0.781	0.967	0.838	0.820	0.906	0.991

MD = Mesio-distal; LL = labio-lingual; IG = incisor-gingival; a = actual across surface measurement.

Z_j in Ω. Denote by R, T the converted images of S1 and S2, respectively (fig. 1). Below we show the result of a pair of such images, first for the molar of one MZ co-twin, S1 (fig. 1) and then for the other, S2 (fig. 1). Here in both cases, plot (a) shows the locations of all (x_i, y_i) in domain Ω, (b) shows locations (X_j, Y_j) we only show a coarse resolution of these points to ensure visibility. Then (c) shows the surfaces as would be observed from Cloud in STL format and (d) shows the same surface as a converted image in Ω. Clearly no essential loss of information has occurred by comparing (c) and (d) visually.

Conversion of Image-Like Data Back to Surface Data after Registration

Once R and T have been obtained in Ω, we use diffusion-based image registration to register T onto R and denote the registered result by T0. Then we use the following procedure to convert the image back to surface data:

(1) Load in the GMESH data $(x_i, y_{i,} z_i)$ and the connectivity table E from molar S1.

(2) Load in the registered image data, denoted by $(X_j, Y_{j,} W_j)$.

(3) Interpolate $(X_j, Y_{j,} W_j)$ onto the new surface positions to obtain all $(x_i, y_{i,} t_i)$ with all surface data represented by t_i. Use E and $(x_i, y_{i,} t_i)$ to build up the registered surface T1 in GMESH format, suitable for further exploration in Cloud.

The 3D measurements included the mesio-distal, labio-lingual and inciso-gingival 2D variables (used for validation of the system against an established 2D system) and the same actual 3D version of those variables across the surface as opposed to between 2 points. In addition the actual surface area was measured. The teeth studied were the upper left centrals and upper right laterals.

Results

Table 1 demonstrates the high degree of Fleiss' Intra-Class Correlation Coefficient (ICCC) [9] found for intra- and inter-operator reliability measurements for the 3D scanning apparatus (Optix 400S scanner, Realscan USB, 3D Digital Corporation, USA). Fleiss' method may be used for inter- as well as intra operator calculations with the addition of one extra 'operator' statistical class. This method accounts for biological variance, and is less forgiving than other methods such as Pearson's Correlation. All data show substantial or excellent reliability [10].

All significance values for both intra-operators were >0.05, the majority being >0.8. This implies no significant difference between measurements.

Figure 2 demonstrates the T1 surface obtained from this registration procedure along with the difference of T1 and S1.

Discussion

The 3D reliability data (table 1) demonstrate the method's good application to this research. Practically all the ICCC reliability values are in the substantial or excellent categories and it should be noted that differences between these categories are statistically insignificant and as tabulated and expected we have slightly lower values for inter-examiner comparisons. However, the upper left central incisor measurement for the LL variable for inter-examiner ICCC data is 0.561 (still moderate). It is clear when handling 3D images that certain variable measurements require greater skill than their 2D counterparts as these images are not fixed and orientation of the images creates its own error, multiplied for inter-examiners. The reader should also remember that, for example, labio-lingual in 2D (between front and rear of the tooth) demonstrating tooth depth is not the same variable in 3D where it includes the full height contour of the tooth.

This initial study has demonstrated an alternate mathematical method for the superimposition of non-identical objects, i.e. the molar crowns of MZ co-twins. The approach has proven to be robust and provides a novel method for deriving a unique output from minimising the least-squares calculations for registration by converting surface data into full 3D voxel linear dataset. The approach is linear and ensures uniqueness, in contrast to the 'best fit' approach which is generally non-linear giving non-convex minimalisation and resulting in the possibility of more than one answer.

Figure 2 demonstrates the T1 surface obtained from this registration procedure along with the difference of T1 and S1: the right hand plot reveals some artefacts generated from the edges of the surfaces, when converting to images. Ideally, it would be useful to smooth out the image edges or redefine the domain Ω. Nevertheless, the current experiments indicate that the central variations between the pairs of surfaces can be compared more accurately with this new Synetic method.

Conclusions

This paper outlines the development of a methodology for enhanced comparison of the dentitions of twins. This will result in a greater degree of discrimination that should provide additional information on the roles of genetic, epigenetic and environmental factors during dental development. This method also should provide additional information from left /right asymmetry within individuals and how this compares to inter MZ twin differences. The initial construction of the 3D data sets has proved reliable when assessed using a wide array of dental variables.

Acknowledgments

This study was partly sponsored by the Wellcome Trust Foundation. Grant number GRO75945MA.

References

1 Proffit WR: On the aetiology of malocclusion. Br J Pediatr Dent 1986;7:155–160.

2 Brook AH, Lath D, Brook BJ, Smith RN: The aetiology of developmental defects of the dentition; in Zadzinska E (ed): Current Trends in Dental Morphology Research. Lodz, University of Lodz Press, 2005, pp 467–474.

3 Dempsey PJ, Townsend GC: Genetic and environmental contributions to variation in human tooth size. Heredity 2001;86:685–693.

4 Townsend GC, Corruccini RS, Richards LC, Brown T: Genetic and environmental determinants of dental occlusal variation in South Australian twins. Aust Orthod J 1988;10:231–235.
5 Su CY, Corby PM, Elliot MA, Studen-Pavlovich DA, Ranalli DN, Rosa B, Wessel J, Schork NJ, Hart TC, Bretz WA: Inheritance of occlusal topography: a twin study. Eur Arch Paediatr Dent 2008;9:19–24.
6 Hill DLG, Batchelor PG, Holden M, Hawkes DJ: Medical image registration. Phys Med Biol 2001;46:1–45.
7 Modersitzki J: Numerical Methods for Image Registration. London, Oxford University Press, 2004.
8 Chumchob N, Chen K: A robust affine image registration method. Int J Numer Anal Mod 2008;in press.
9 Fleiss JL: Reliability of Measurement: The Design and Analysis of Clinical Experiments. New York, Wiley, 1986, vol 1, pp 1–32.
10 Donner A, Eliasziw M: Sample Size Requirements for Reliability Studies. Stat Med 1987;6:441–448.

Dr. Richard N. Smith
International Collaborating Centre for Oro-facial Genetics and Development, School of Dental Sciences
Edwards Building, University of Liverpool
Liverpool, L69 3GN (UK)
Tel. +44 151 706 5118, Fax +44 151 706 5857, E-Mail R.N.Smith@liv.ac.uk

Koppe T, Meyer G, Alt KW (eds): Comparative Dental Morphology.
Front Oral Biol. Basel, Karger, 2009, vol 13, pp 148–152

A Three-Dimensional Comparison of the Modifying Effects of Familial Genetic Contribution in Turner Syndrome

L.R. Horrocks[a] · A. Brook[a] · L. Alvesalo[a,b] · R.N. Smith[a]

[a]International Collaborating Centre for Oro-Facial Genetics and Development, School of Dental Sciences, University of Liverpool, Liverpool, UK; [b]Institute of Dentistry, University of Oulu, Oulu, Finland

Abstract

Comparison of individuals affected by chromosomal aneuploidies with controls has already confirmed the differential action of X and Y chromosomes during dental ontogeny. Permanent enamel and dentin structures are formed during the development of the dentition and these reflect the combined effects of both genetic and environmental factors during the prenatal and early postnatal period. It can provide key information on the effect of developmental assaults such as chromosomal aneuploidy as it is not subject to subsequent resorption or remodelling like other skeletal structures within the body. Dental features seen in Turner syndrome (TS) include significantly smaller teeth than controls and a narrow, deeply vaulted palate. Familial genetic contribution is known to have a modifying effect on features such as palate width. During the last two decades imaging and analysis techniques have undergone rapid evolution. Each development offers new possibilities for describing and defining structures within the body. This pilot study used a novel 3D system to measure the dentition of 10 individuals with TS and 10 unrelated female controls. The reliability of the method and its effectiveness in quantitatively describing differences between the two groups was verified. A larger study involving TS individuals, their first degree unaffected female relatives and female controls (n = 35 each group) is underway which will use the new 3D system to describe the surface morphology, volume and contours of the dentition in order to quantify the effect of TS on dental development and to assess any modifying effects of familial genetic contribution.

Advances in developmental biology have greatly enhanced our understanding of the physiological and genetic processes involved in dental development [3]. The genes involved in dental development are highly conserved and therefore, inferences drawn from one species can be applied to another. It is relatively easy to manipulate the genetic makeup of study species such as the mouse and the effect on the resultant offspring can be examined to provide insight into the relative contribution of different factors to the overall development process. Conclusions drawn from such experiments can then form the basis for less invasive studies in humans who exhibit naturally occurring genetic anomalies. Permanent enamel and dentin structures are formed during the development of the dentition and these reflect the combined effects of both genetic and environmental factors during the prenatal and early postnatal period. They are not subject to subsequent resorption or remodelling like other skeletal structures within the body and therefore provide key information about the timing of developmental defects. The differential action of X and Y chromosomes on tooth crown size has already been confirmed in studies involving individuals with chromosome anomalies [1].

Alvesalo's KVANTTI collection (Institute of Dentistry, Oulu University, Finland) contains a large number of dental casts from individuals with sex chromosome anomalies such as Turner syndrome (TS). Individuals with TS are females with only one fully functional X chromosome rather than the normal complement of two. There are several different karyotypes; the most common being 45X0 which affects approximately 50% of TS subjects. A further 15% of TS have two X chromosomes of which one is structurally abnormal. The remainder of cases are mosaic forms in which only some body cells are affected and the rest are normal. Symptoms vary according to karyotype with the mosaic form least affected and 45X0 most severe [4]. The most commonly occurring symptoms are reduced height and gonadal dysgenesis. TS individuals have significantly smaller teeth than controls and the palate is narrower and often deeply vaulted [5].

Variations in the number or structure of sex chromosomes are a relatively common type of chromosomal anomaly with a frequency of 1:500 live births. The associated symptoms are more compatible with survival than autosomal disorders because sex chromosomes contain a relatively high level of facultatively inactive heterochromatin and less transcriptively active genes than the other chromosomes [6]. Turner syndrome (TS) affects approximately 1:2,000 live female births although pre-natal incidence is estimated to be much higher as the majority of foetuses are lost during development [5]. Lyonisation [7] was thought to result in the inactivation of most genes on the second X chromosome in females but one which escapes inactivation is the SHOX gene [8] which is expressed in the limb buds and the first and second pharyngeal arches during development [9]. Although its full effects are not yet known the short stature associated with TS is thought to be caused, in part, by haploinsufficiency of SHOX. Individuals with a SHOX deletion which is proximal to the junction between Xp22.2 and Xp22.3 are also classified as having TS [10]. The amelogenin gene is located at Xp22 and haploinsufficiency leads to reduced enamel thickness in TS [11]. Palate width has been shown to increase with sex chromosome number [12] and the characteristic high arched narrow palate in TS may be associated with an increased incidence of cleft palate [13]. It is worth noting that although many TS patients receive treatment with growth hormones and oestrogen it has been shown that neither has a significant effect on dental development [14].

Rapid improvement in imaging and analysis techniques during the last two decades offers new possibilities for describing and defining structures within the body. Measurements with manual or 2D methods such as callipers or computer analysis of photographs have provided significant insights into the effect of chromosomal abnormalities on tooth size but are limited in the amount of descriptive information that they can supply. This study used 3D laser profilometry together with novel image analysis software to measure the dentition of 10 individuals with TS and 10 unrelated female controls. In comparison to earlier methods the new 3D system facilitates many additional measurements which represent the true surface contour of the dentition allowing its morphology to be described in a more sensitive manner than previously possible. The aim was to test the reliability of the method and its effectiveness in quantitatively describing differences between two groups in preparation for a larger study.

Materials and Methods

A pilot study consisting of the upper casts of ten 45X0 Turner syndrome (TS) whose karyotype had been confirmed for medical reasons and ten unaffected and unrelated female controls were selected from the KVANTTI collection. Copies of the original dental casts were taken using an impression machine with 2-mm EVA discs (Erkodent, Germany) and cast using hard dental stone (Kaffir D, British Gypsum). The study received ethical approval from the University of Liverpool Research Ethics Committee (RETH000075).

The study models were digitized with an Optix 400S scanner (Realscan USB, 3D Digital Corporation, USA) at a resolution of 750 points ×1,000 lines per mm. A Realscan rotary table allowed automated 360° scans to be taken at pre-set intervals. The resultant files were imported into SLIM 3D software where the individual images were merged into a complete 3D model which was then transferred to CLOUD software for final processing and analysis. In order to validate reliability of the 3D system with an established 2D method, digital vernier callipers (Draper Tools, UK model 52427) were used to obtain comparative mesial-distal (MD) and occlusal-gingival (OG) measurements of four teeth: right central, left lateral, right 1st premolar and left 1st molar. The 3D system was used to obtain projected (straight line) MD and OG measurements of the same teeth as these could be directly compared with standard 2D ones. In order to assess the ability of the 3D system to describe differences between the groups, additional novel measurement variables (actual perimeter (P) and actual surface area (SA)) were obtained. Intra-operator reliability was assessed by producing 10 duplicate scans and comparing the measurements taken.

For each variable, limits of agreement [15] were calculated within which 95% of the measurement differences were expected to lie in order to assess whether any bias was present. Pearson's correlation, a paired t test and Fleiss' intra-class correlation coefficients (ICCC) were used to assess whether there was any significant difference between the two sets of measurements. The ability of the new 3D system to quantitatively describe differences in morphology between TS and controls was assessed by comparing the TS and control group. A Levene test was applied to confirm whether variances were equal and an independent t test to assess whether the groups were significantly different. Average mean size for each variable was compared between the groups and the results compared with previously published TS literature.

Results

2D/3D Method Correlation

Bland Altman plots showed that no systematic bias was present. A paired t test showed no significant difference between the two methods ($p \geq 0.05$) and Pearson's product moment correlation co-efficient confirmed that correlation between measurements with digital callipers and the laser scanning system was highly significant ($r > 0.99$, $p \leq 0.01$). Fleiss' ICCC (>0.99) showed excellent reliability according to the classification set down by Donner and Eliasziw [16].

Intra-Operator Reliability for 3D Method

Bland-Altman plots showed no systematic bias was present and Fleiss' ICCC (>0.98) showed excellent agreement between measurements of repeated scans [16].

Comparison of 45X0 Turner Group and Controls

A Levene's test performed on data from the 45X0 and controls showed that variances were equal between groups and the data did not require transformation ($p > 0.05$). The independent-measures t test confirmed that the groups were significantly different from each other ($p < 0.05$) and comparison of the means for each variable showed that tooth dimensions of controls were larger than the 45X0 in all cases. Table 1 shows descriptive statistics for the upper right central incisor.

Discussion

Measurements with the 3D laser scanning and analysis system showed excellent agreement with those from an established 2D method according to Donner and Eliasziw's [16] classification of Fleiss' ICCC. Intra-operator reliability was also excellent. Comparison of groups showed that tooth dimensions of controls were significantly larger than TS for all variables which agreed with previously published literature [17]. In addition to being a reliable and effective method for quantifying differences in morphological traits in dentition between TS and control groups the 3D system provided scope for the creation of new variables which follow the actual surface

Table 1. Comparison of 45X0 Turner syndrome group (n = 10) and controls (n = 10) – descriptive statistics for UR1

Variable	OG	MD	P	SA
Mean difference (control – TS)	1.01	1.05	4.14	29.09
SE mean difference	0.38	0.17	0.73	4.82
Independent t test (p value)	0.02	0.01	0.01	0.01

morphology of a tooth rather than a projected straight line value between two points. Accurate positioning of markers is facilitated by the ability to rotate the 360° images on screen during analysis. Digital archiving of computerised 3D images of dental casts provides a safe and space efficient way of storing information which can then be accessed repeatedly to form the basis of future studies and allows electronically transfer of images between research institutes thus facilitating collaboration.

A correlation has already been shown between the final height of mothers and their offspring in the general population and a similar correlation is seen between parents and their offspring with TS [18]. This relationship has also been seen in some craniological features [19] and palate width [2]. In the majority of cases X chromosome monosomy results from non-disjunction during paternal meiosis meaning that in 70–80% of TS cases the sole X chromosome is of maternal origin [20]. A larger study comprising of 105 casts (upper and lower) evenly split between 45X0, closely related 46XX females and unrelated 46XX controls is underway which will use the 3D system to obtain a range of projected and actual measurements to quantitatively describe differences in the dental morphology present between groups and to ascertain whether familial genetic contribution has a modifying effect on the features observed.

Conclusions

Limits of agreement, Fleiss, Pearsons and t tests validate the reliability and effectiveness of the 3D laser profiler and computer analysis system to quantify differences observed in dental morphology between groups. Results obtained with the 3D system show significantly reduced tooth size in TS compared to normal 46XX controls. Novel 3D variables which follow the actual surface of a tooth rather than a projected A-B straight line measurement offer the prospect of gaining increased insight into the effect of TS on dentition.

References

1 Alvesalo L: Sex chromosomes and human growth. Hum Genet 1997;101:1–5.
2 Perkiomaki MR, Alvesalo L: Palatine ridges and tongue position in Turner syndrome subjects. Eur J Orthodont 2008;30:163–168.
3 Butler P.M: Ontogenetic aspects of dental evolution. Int J Dev Biol 1995;39:25–34.
4 Varrela J, Vinkka H, Alvesalo L: The phenotype of 45X females: an anthropometric quantification. Ann Hum Biol 1984;11:53–66.
5 Gravholt CH: Epidemiological, endocrine and metabolic features in Turner syndrome. Eur J Endocrinol 2004;151:657–687.

6 Chadwick BP, Willard HF: Multiple spatially distinct types of facultative heterochromatin on the human inactive X chromosome. Proc Natl Acad Sci 2004;101:17450–17455.
7 Lyon MF: Some milestones in the history of X-chromosome inactivation. Annu Rev Genet 1992;26:17–28.
8 Rao E, Weiss B, Fukami M, Rump A, Niesler B, Mertz A, Muroya K, Binder G, Kirsch S, Winkleman M, Nordsiek G, Heinrich U, Breuning M, Ranke MB, Rosenthal A, Ogata T, Rappold GA: Pseudoautosomal deletions encompassing a novel homebox gene cause growth failure in idiopathic short stature and Turner syndrome. Nat Genet 1997;16:54–63.
9 Clement-Jones M, Schiller S, Rao E, Blaschke RJ, Zuniga A, Zeller R, Robson SC, Binder G, Glass I, Strachan T, Lindsay S, Rappold GA: The short stature homebox gene SHOX is involved in skeletal abnormalities in Turner syndrome. Hum Mol Genet 2000;9:695–702.
10 Kosho T, Muroya K, Nagai T, Fujimoto M, Yokoya S, Sakamoto H, Hirano T, Terasaki H, Ohashi H, Nishimura G, Sato S, Matsuo N, Ogata T: Skeletal features and growth patterns in 14 patients with haploinsufficiency of SHOX: implications for the development of Turner syndrome. J Clin Endocrinol Metab 1999;84:4613–4621.
11 Lau EC, Mohandras TK, Shapiro LJ, Slavkin HC, Snead ML: Human and mouse amelogenin gene loci are on the sex chromosomes. Genomics 1989;4:162–168.
12 Laine T, Alvesalo L: Palatal and mandibular arch morphology in 47XXY men and in other sex-chromosome anomalies. Arch Oral Biol 1993;38:101–105.
13 Gorlin RJ, Cohen MM, Henneken RCM: Chromosomal syndromes: common and/or well known syndromes. Turner syndrome; in: Syndromes of the Head and Neck, ed 4. New York, Oxford University Press, 2001.
14 Hass AD, Simmons KE, Davenport ML, Proffitt WR: The effect of growth hormone on craniofacial growth and dental maturation in Turner syndrome. Angle Orthod 2001;71:50–59.
15 Bland J, Altman DJ: Statistical method for assessing agreement: two methods of clinical measurement. Lancet 1986;i:307–310.
16 Donner A, Eliasziw M: Sample size requirements for reliability studies. Stat Med 1987;6:441–448.
17 Townsend GC, Jensen BL, Alvesalo L: Reduced tooth size in 45X (Turner syndrome) females. Am J Phys Anthropol 1984;65:367–372.
18 Brook CG, Gasser T, Werder EA, Prader A, Vanderschueren-Lodewykx MA: Height correlations between parents and mature offspring in normal subjects and in subjects with Turner's and Klinefelter's and other syndromes. Ann Human Biol 1977;4:17–22.
19 Perkiomaki MR, Kyranides S, Niinimaa A, Alvesalo L: The relationship of distinct craniofacial features between Turner syndrome females and their parents. Eur J Orthodont 2005;27:48–52.
20 Sagi L, Zuckerman-Levin N, Gawlik A, Ghizzoni L, Buyukgebiz A, Rakover Y, Bistritzer T, Admoni O, Vottero A, Baruch O, Fares F, Malecka-Tendera E, Hochberg Z: Clinical significance of the parental origin of the X chromosome in Turner syndrome. J Clin Endocrinol Metab 2007;92:846–852.

Linda Horrocks, BSc
International Collaborating Centre for Oro-facial Genetics and Development, School of Dental Sciences
Edwards Building, University of Liverpool
Liverpool, L69 3GN (UK)
Tel. +44 151 706 5665, Fax +44 151 706 5857, E-Mail L.R.Saunders@liverpool.ac.uk

Koppe T, Meyer G, Alt KW (eds): Comparative Dental Morphology.
Front Oral Biol. Basel, Karger, 2009, vol 13, pp 153–157

Permanent Tooth Formation as a Method of Estimating Age

Helen M. Liversidge

Barts and The London School of Medicine and Dentistry, London, UK

Abstract

The aims of this study were to describe maturity data of permanent tooth formation from a large sample and to adapt these for estimating age, to compare mean age dental formation stages between ethnic groups and sexes. **Methods:** This was a retrospective, cross-sectional study of 1,050 panoramic radiographs of healthy dental patients in London aged 2 to 22. Similar numbers of each sex and ethnic group (White and Bangladeshi) were selected for each year of age. Permanent mandibular teeth were scored using 14 stages described by Moorrees and co-workers in 1963 plus crypt stage. Mean age of each tooth stage (age when 50% of sample had reached/passed each stage) was calculated using probit regression for males and females by ethnic group for each stage. Data were combined where no significant difference between mean age for groups was observed. Average age 'within a stage' was also calculated for each tooth stage. **Results:** No ethnic difference was noted in mean age. Canine root stages and third molar apex stages were significantly different between the sexes. The average age of most stages was considerably later than that given by Moorrees and co-workers. Accuracy of age estimation on a separate test sample of radiographs was considerably more accurate using these new data. **Conclusions:** These results provide an accurate method to estimate age from developing permanent mandibular teeth. The lack of ethnic difference in dental maturity of individual teeth, suggest that these findings might be appropriate to accurately estimate age for other groups.

Developing teeth are often used to estimate age, but few dental developmental standards encompass the entire growth span. The aim of this study was to present timing of permanent tooth stages from a large, wide age range radiographic sample and present these data adapted for estimating age. In addition, mean age of attainment of permanent tooth formation times was compared between males and females and between two ethnic groups in London, UK.

Material and Methods

The sample studied was archived panoramic radiographs of 1,050 healthy dental patients in London, made up of similar numbers of two ethnic groups (white and Bangladeshi local residents) for each year of age from 2 to 22. Radiographs from 25 males and 25 females were selected for each year of age with the exception of two year olds where the numbers were 29 male and 21 female. The same number of individuals per year of age was selected to improve accuracy (standard deviation) of estimating age for younger and older individuals.

All radiographs had been taken in the course of diagnosis and treatment. All permanent teeth were scored using 14 stages described by Moorrees et al. [1] plus crypt stage. These include cusp tip initiation, coalescence of cusp tips, cusp outline complete, crown fractions, crown complete, initial root, root cleft formation, root fractions,

Table 1. Mean age entering a formation stage, SD and adapted mean for age estimation

Stage	I1		I2		C		P1		P2		M1		M2		M3	
Mean age and SD of entering a tooth formation stages (n = 1,050)																
Cr									2.72	0.97			2.44	1.31	8.65	2.32
Ci									3.34	0.82			3.34	0.87	9.67	2.03
Cco									3.81	0.80			4.01	0.83	10.18	1.64
Coc							3.40	0.82	4.49	0.90			5.02	0.91	11.06	1.58
C1/2							3.88	0.76	5.06	0.94			5.51	1.00	11.60	1.41
C3/4			2.72	0.71	4.20	0.82	5.24	0.94	6.04	0.87			6.54	0.84	13.09	1.45
Cc	3.30	0.69	3.89	0.72	5.27	0.74	6.14	0.85	6.70	0.76	3.20	0.58	7.40	0.85	13.89	1.43
Ri	3.99	0.77	4.78	0.81	5.97	0.84	6.94	0.79	7.59	0.98	3.60	0.69	8.07	0.95	14.68	1.80
Rcl											4.00	0.78	8.59	0.90	14.88	1.43
R1/4	5.07	0.76	5.69	0.81	6.82	0.78	7.88	0.99	8.59	0.99	4.80	0.83	9.29	0.92	15.50	1.68
R1/2	6.24	0.83	6.67	0.67	8.14	0.96	8.92	0.93	9.44	1.10	5.77	0.91	10.47	1.14	16.69	1.93
R3/4	6.70	0.68	7.09	0.73	9.15	1.07	9.88	1.22	10.23	1.16	6.85	0.60	11.80	1.21	17.54	1.94
Rc	7.28	0.84	7.65	0.73	10.89	1.56	11.54	1.44	12.24	1.23	7.54	0.87	12.85	1.19	18.35	1.96
A1/2	7.55	0.76	8.08	0.99	11.61	1.47	12.06	1.38	12.76	1.08	8.45	1.17	13.63	1.22	19.13	2.10
Ac	8.04	0.90	8.69	1.07	12.23	1.42	12.70	1.25	13.26	1.21	9.38	1.23	14.52	1.21	20.23	2.33
Adapted mean for estimating age[1]																
Cr									3.03				2.89		9.16	
Ci									3.52				3.68		9.92	
Cco									4.15				4.52		10.62	
Coc							3.64		4.78				5.26		11.33	
C1/2							4.56		5.55				6.02		12.34	
C3/4			3.30		4.73		5.69		6.37				6.97		13.49	
Cc	3.64		4.34		5.61		6.54		7.14		3.42		7.74		14.28	
Ri	4.53		5.23		6.38		7.41		8.09		3.79		8.33		14.78	
Rcl											4.38		8.94		15.19	
R1/4	5.65		6.18		7.47		8.40		9.02		5.22		9.88		16.10	
R1/2	6.47		6.88		8.63		9.40		9.84		6.26		11.14		17.12	
R3/4	6.99		7.33		10.03		10.71		11.24		7.21		12.33		17.94	
Rc	7.41		7.86		11.26		11.80		12.50		8.00		13.24		18.74	
A1/2	7.80		8.38		11.92		12.38		13.01		8.92		14.08		19.68	

[1] Adapted by adding half the interval between stages.

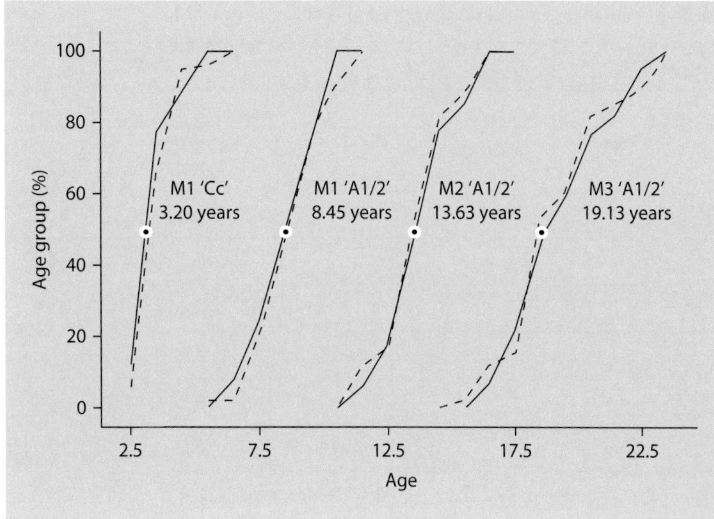

Fig. 1. Comparison of cumulative curves showing similarity of mean age of ethnic groups for M1 stage 'Cc' and 'A1/2', M2 stage 'A1/2' and M3 stage 'A1/2'. Solid line = Percentage of White London children; dashed line = percentage of Bangaldeshi London children for age group.

root length complete, apex half closed and apex closed (radiographically). Some stage descriptions for molars and intra observer variation are detailed elsewhere [2]. Mean age of each dental maturational event (age when 50% of sample had reached/passed each stage) of each tooth was calculated using probit regression for males, females and ethnic group using one year age groups. Mean age was adapted for estimating age by adding half the interval to the next stage [3], with the exception of 'apex closed' stage as once a tooth apex matures, it can no longer be used to estimate age using this method. Average age 'within a stage' was also calculated.

Results

No ethnic differences in mean age were observed. The similarity between ethnic groups is illustrated in figure 1, where the proportion of children in some molar stages is plotted against age. This figure also shows the wide age range for these stages. The steep slope of the line for 'crown complete' of the first permanent molar (M1) reflects a small age range, while the range for 'apex half closed' for the third permanent molar is from 15 to 22 years. Most comparisons (85 of 97) of mean age were not significantly different between males and females. Exceptions were the latter half of root stages of the canine, apex closure of the first molar and apical stages of the third molar. Of these, mean age in females was earlier than males except for third molar stages. Mean age and standard deviation for combined groups and adapted mean age for estimating age are presented in table 1. Average age 'within a stage' (plus standard deviation and N) are shown in table 2.

Discussion

The mean ages of many teeth from the present study are considerably later than some previous standards [1, 4–6], although these early studies do not give details of standard error of mean age. Mean age in the present study was more than a year later in almost a third of comparison with Moorrees et al. [1]. Consequently, estimating age of London children using maturity data from Moorrees et al. [1] is likely to underestimate age considerably. This was tested using the test sample of Maber et al. [7] and accuracy (bias) using Moorrees et al. [1] was more than 1 year (under

Table 2. Mean age 'within a stage', SD and N

Stage	I1		I2		C		P1		P2		M1		M2		M3	
Crypt							2.60	0.37	3.23	0.36			3.04	0.37	8.78	1.65
								8		26				29		49
Ci							2.87	0.35	3.72	0.63			3.91	0.76	8.83	0.91
								16		24				37		28
Cco							3.09	0.41	4.16	0.89			4.51	0.74	10.43	1.48
								41		31				46		43
Coc					2.96	1.29	3.65	0.54	4.89	0.74			5.53	0.97	10.87	1.10
						18		24		29				24		26
C1/2	2.76	0.67	2.79	0.32	4.00	0.69	4.70	0.86	5.44	0.72	2.67	0.26	5.76	0.71	12.28	1.55
		7		34		87		67		51		22		52		75
C3/4	3.09	0.54	3.53	0.59	4.76	0.69	5.58	0.75	6.20	0.81	2.98	0.36	6.97	0.86	13.29	1.58
		56		48		55		46		33		38		44		38
Cc	3.91	0.59	4.48	0.69	5.64	0.64	6.39	0.93	7.21	1.02	3.55	0.50	7.79	1.07	14.29	1.57
		25		43		35		39		42		20		32		27
Ri	4.54	0.69	5.17	0.69	6.30	0.89	7.46	0.90	8.10	0.90	3.95	0.53	8.20	0.94	15.15	1.77
		54		46		43		47		50		21		26		26
Rcl											4.51	0.74	8.92	0.81	15.48	1.40
												39		36		24
R1/4	5.64	0.78	6.02	0.82	7.46	0.96	8.34	0.91	9.02	1.11	5.27	0.93	10.01	1.09	16.46	1.87
		58		46		63		51		42		47		59		57
R1/2	6.19	0.73	6.75	0.58	8.73	1.08	9.68	1.05	10.02	1.08	6.10	0.74	11.20	1.25	17.11	2.00
		22		23		51		48		40		56		66		41
R3/4	7.04	0.70	7.33	0.62	10.37	1.32	10.89	1.42	11.32	1.33	7.50	1.18	12.29	1.13	17.94	1.92
		30		28		88		84		105		35		55		39
Rc	7.17	0.56	8.28	0.74	11.12	1.30	11.61	1.24	12.12	1.18	8.24	1.11	13.49	1.37	19.02	1.94
		14		21		37		27		25		44		40		37
A1/2	7.95	0.67	8.52	0.89	11.88	1.47	12.16	1.14	13.32	1.23	8.94	1.14	13.90	1.33	19.67	1.62
		25		30		31		33		23		46		45		50

estimating age by 1.19 year) with median absolute difference of 1 year. Accuracy of Smith's [3] adaptation fared considerably better with bias of 0.67 years (underestimating age) and median absolute difference of 0.64 years. Accuracy of the adapted data from table 1 on the large test sample was considerably better at 5 weeks (0.10 years) with a median absolute difference of around 6 months (0.55 years). Table 2 was marginally more accurate at 2 weeks (0.04 years) with a similar median absolute difference (0.53 years).

There are several reasons why results from the present study differ to previous dental maturity studies. The Fels/Forsyth collaboration [1, 4–6] was unique in that individuals were X-rayed from birth and remain the major studies reporting early tooth formation from radiographs. It is no longer considered ethical to X-ray growing children without a diagnostic reason. The minimum age of the present study was 2 years from archived panoramic radiographs of children with caries. Another explanation is assessment or scoring of tooth stages; particularly fractions of crown height or root length. Secular trend in the timing of tooth formation is a possible explanation but seems unlikely as more detailed results become available from histological studies showing similarities between the past and the present and between different world groups.

The variation in the timing of individual tooth maturation is large for most stages of tooth formation and increases with age. Radiographic evidence of M2 crypt was seen in the present study from age 2 to 4 while the M3 crypt could be as early as age four or as late as eleven. Standard deviation for most stages was from just less than a year to just over 2 years. This has implications when using dental maturity to estimate age as the 95% CI for an individual is likely to be between 3 and 4.5 years.

Conclusions

These finding present a new method to estimate age more accurately from individual developing mandibular permanent teeth. The lack of ethnic difference in average age of tooth formation stages between whites and Bangladeshis in London suggests that this method of estimating age might be appropriate to other groups.

References

1 Moorrees CFA, Fanning EA, Hunt EE: Age variation of formation stages for ten permanent teeth. J Dent Res 1963;42:1490–1502.
2 Liversidge HM: Timing of human mandibular third molar formation. Ann Hum Biol 2008;35:294–321.
3 Smith BH: Standards of human tooth formation and dental age assessment; in Kelley M, Larsen CS (eds): Advances in Dental Anthropology. New York, Liss, 1991, pp 143–168.
4 Fanning EA, Brown T: Primary and permanent tooth development. Aust Dent J 1971;16:41–43.
5 Fanning EA: A longitudinal study of tooth formation and root resorption. NZ Dent J 1961;57:202–217.
6 Gleiser I, Hunt E: The permanent mandibular first molar: its calcification, eruption and decay. Am J Phys Anthropol 1955;13:253–283.
7 Maber M, Liversidge HM, Hector MP: Accuracy of age estimation of radiographic methods using developing teeth. For Sci Int 2006;159;68–73.

Dr. Helen M. Liversidge
Queen Mary University of London
Barts and The London School of Medicine and Dentistry
Turner Street, London, E1 2AD (UK)
Tel. +44 207 882 8619, Fax +44 207 377 2174, E-Mail h.m.liversidge@qmul.ac.uk

Koppe T, Meyer G, Alt KW (eds): Comparative Dental Morphology.
Front Oral Biol. Basel, Karger, 2009, vol 13, pp 158–161

Teeth and Reconstruction of the Past: An Introduction

John R. Lukacs

Department of Anthropology, University of Oregon, Eugene, Oreg., USA

The role of teeth as tools in reconstructing the nature and behavior of past populations is diverse and widely appreciated. This appreciation is shared by scientists and by the educated public who understand the significance of intergroup differences in dental disease, tooth size, wear patterns, crown and root morphology, and incremental microstructures of dental tissues for deciphering the past. In the broadest sense, dental anthropologists employ phenotypic variation in the dentition as a research tool to evaluate the dynamic and interactive impact of physical and cultural environments on human biology [1, 2]. The topical breadth of analytic problems embraced within this research paradigm are eclipsed only by the diverse methodologies adopted by investigators.

For the first time in its history, the International Symposium on Dental Morphology in Greifswald included a session dedicated to bioarchaeological perspectives derived from teeth. The diversity of podium and poster presentations illustrate well the wide range of research problems and methods in this research paradigm. A sample of this topical diversity is briefly presented in this introduction to exemplify the diverse ways in which symposium participants illustrated the use of teeth as tools in reconstructing the past.

The session began with a sumary review of the value of teeth in reconstructing the past, with examples from Çatalhöyük, Turkey, by Simon Hilson. Appropriately, dental morphology was the most common theme in this session comprising the subject of five presentations. Patterns of morphological variation in permanent or deciduous teeth were used to assess population structure or inter-group biological affinity among the ancient Maya of Yucata (Cucina and Wrobel), Bell Beaker populations of Europe (Desideri), and modern Javanese Malay of Indonesia (Lukacs and Kuswandari). The frequency of C-shaped root morphology revealed continuity between modern and pre-Hispanic groups of Yucatan (Cucina and associates). Variation in tooth morphology was creatively combined with other indicators of individual identity derived from dental tissues (aDNA, stable isotope and elemental composition) to track migration patterns in prehistoric Thailand (Cox and colleagues) and to illustrate the importance of 'dental fingerprinting' in reconstructing early population history (Krantzbühler and Alt). Dental pathology and tooth wear were each represented by three presentations. Caries diagnosis, etiology (Lukacs) and prevalence through periods of subsistence transition (Jackes; Tayles and colleagues) were discussed, yielding important cautionary lessons for consideration as research on dental caries and subsistence continues. Contributions on tooth wear analyzed Inuit-Neanderthal differences (Clement), the adaptiveness of edge-to-edge bite (Rossbach and Alt), and the relationship between tooth size and tooth wear (Vodanovi). Incremental markers of dental enamel and cementum were discussed in three papers; one focused on childhood growth in early London (Antoine and colleagues) and another on an ancient infant cemetery in Greece (FitzGerald and associates). A podium presentation on the cultural treatment of teeth focused on dental ablation in sub-Saharan Africa and included a significant ethnographic perspective, while directional and fluctuating asymmetry from different historical samples (Gawlikowska-Sroka and colleagues) comprised the subject of an informative poster presentation. Unfortunately this topical and methodological diversity could not be fully presented in this volume due to the publishers guidelines, which included strict page and chapter limitations. Nevertheless, some feeling for the substance of the session 'Teeth and the past' can be derived from the five papers appearing in this section. Below I explain the intellectual foundation for this section's organization and structure, then summarily introduce each paper.

Symposium organizers asked session chairs to invite two speakers. Intrigued by continuing issues associated with reconstructing past dietary behavior from the dentition, I decided to focus invited speakers' attention on problems such as accurate diagnosis, differences in prevalence and diverse etiologies of dental caries. Consequently, I selected two investigators well qualified to address the relationship between diet and dental pathology in prehistoric populations. A problem of longstanding is that the prevalence of dental caries in past populations is often employed as a direct indicator of the degree of reliance on agriculture. The decline in dental health with the rise of civilization and especially intensive agriculture has been appreciated for more than half a century [3]. More than two decades have passed since the increase in caries prevalence was directly associated with the intensification of maize agriculture among Native North Americans [4]. Another decade or more has passed since initial reconstructions of prehistoric diet were initially inferred from dental caries prevalence in populations 'on the margins' of the Old World: Iberia and Southeast Asia.

The first two papers in this section focus on dental caries, and more narrowly on methodology and subsistence concerns. Nancy Tayles and colleagues (University of Otago, Dunedin) ask if dental caries rates constitute sufficient evidence for farming in prehistoric Asia, while Mary Jackes (University of Waterloo) addresses methodologic, taphonomic, and demographic factors that complicate the ability to accurately infer subsistence change from changes in caries rates in Portugal. The impact of rice cultivation on oral pathology in Southeast Asia was first addressed by Tayles and co-workers, yielding a relationship at variance with the data derived from Native North-American maize cultivators [5]. Methodological issues of recognizing and precisely recording dental caries lesions in populations spanning the Mesolithic-Neolithic transition in Portugal were initially addressed by Jackes and colleagues (University of Waterloo) [6, 7]. The invited speakers were requested to reassess their initial findings on the relationship caries and diet in prehistory in their respective study regions in light of recent developments in the field. These papers underscore just how inappropriate and misleading it can be to generalize pathology and subsistence patterns derived from one region (say prehistoric North America), people, environment and cultigen to another. They also highlight how important it is to have a rigorous and high level of 'quality control' in place (understanding of sample recovery, taphonomy, inter-, and intra-observer variance and demography) before making estimates or comparisons of dental caries prevalence.

Regrettably, health issues prevented Dr. Tayles from attending the symposium, and in consultation with Dr. Koppe, I agreed to fill Dr. Tayles' place in the program with a thematically relevant talk entitled 'Fertility enhances the impact of agriculture on women's caries rates'. An expanded version of this lecture recently appeared in *Current Anthropology* [8]. It reported results of a recently completed a meta-analysis of sex differences in dental caries experience with age and parity in a living population, in an effort to better interpret changes in caries prevalence in past populations – especially at the onset of farming. The increase in fertility that accompanies the onset of sedentary agricultural is interpreted to contribute to the significant increase in women's caries experience, compared with men's, across this critical subsistence transition.

An analysis of dental caries prevalence in historical lowland Polish skeletal samples by Kurek and colleagues provides new data for this region and completes the discussion of dental caries and the past. The possible influence of malnutrition on caries rates in the Kujawy sample is raised and recent research on the diminished anti-microbial capacity of saliva in malnourished children provides a potential mechanism [9].

Two chapters focus on inferences about the past that can be derived from the analysis of dental microstructures of enamel and cementum. In the first of these, FitzGerald and Hillson provide a preliminary report of progress on the histological analysis of deciduous tooth growth in a large sample of infants from the Greek cemetery of Kylindra in the Dodecanese Archipelago. This study extends Fitzgerald's meticulous analysis of dental development, age assessment and health of infants to a new area and promises to ultimately yield new growth standards for long bone, skull and tooth development [10, 11]. By contrast, methods for quantifying cementum annulations hold potential for estimating age at death in archaeological samples [12, 13]. In the next chapter, Obertova and Francken, present a methodological assessment testing the use of cementum annulations to estimate age at death in a recent, known age and sex, sample of teeth from Germany and Sri Lanka. In the final chapter in this section, Alt and Rossbach address the inter-relationship between occlusal variation and dental wear in chronologically and regionally diverse samples from Germany. The inclusion of clinical and anthropological perspectives on the development of edge-to-edge bite, reveals that age-dependent nature of this natural and non-pathological process of dental wear.

While these five chapters provide some valuable lessons in the use of teeth as tools in reconstructing the past, they represent a small subset of the significantly more diverse subjects and methods presented in this session at the symposium in Greifswald.

References

1 Perzigian AJ: Teeth as tools for prehistoric studies. Southern Anthropol Soc Proc No 11, pp 101–114.

2 Goodman AH: Teeth as tools: dental development defects as records of dietary insufficiency; in Kennedy, Brenda V, LeMoine, Genevieve M (eds): Diet and Subsistence: Current Archaeological Perspectives. Calgary, University of Calgary Archaeological Association, 1988, pp 260–266.

3 Klatsky M, Fischer RL: The Human Masticatory Apparatus: An Introduction to Dental Anthropology. Brooklyn, Dental Items of Interest Publ Co, 1953.

4 Cohen M, Armelagos G: Paleopathology at the Origins of Agriculture. Orlando, Academic Press, 1984.

5 Tayles N, Domett K, Nelsen K: Agriculture and dental caries? The case of rice in prehistoric southeast Asia. World Archaeol 2000;32:68–83.

6 Jackes M, Lubell D: Dental pathology and diet: second thoughts; in Otte M (ed): Nature et Culture: Actes du Colloque International de Liége. Liége, Études et Recherches Archéologiques de L'Université de Liége, 1996, pp 457–480.

7 Jackes M, Lubell D, Meiklejohn C: Healthy but mortal: human biology and the first farmers of western Europe. Antiquity 1997;71:639–658.

8 Lukacs JR: Fertility and agriculture accentuate sex differences in dental caries rates. Curr Anthropol 2008;49:901–914.
9 Psoter WJ, Reid BC, Katz RV: Malnutrition and dental caries: a review of the literature. Caries Res 2005;39:441–447.
10 Fitzgerald C, Saunders S, Bondioli L, Macchiarelli R: Health of infants in an Imperial Roman skeletal sample: perspective from dental microstructure. Am J Phys Anthropol 2006;130:179–189.
11 FitzGerald C , Rose J: Reading between the lines: dental development and subadult age assessment using microstructural growth markers of teeth; in Katzenberg M, Saunders S (eds): The Biological Anthropology of the Human Skeleton. New York, Wiley, 2008, pp 163–186.
12 Wittwer-Backofen U, Buckberry J, Czarnetzki A, et al: Basics in paleodemography: A comparison of age indicators applied to the early Medieval skeletal sample of Lauchheim. Am J Phys Anthropol 2008;137:384–396.
13 Wittwer-Backofen U, Gampe J, Vaupel JW: Tooth cementum annulation for age estimation: results from a large known-age validation study. Am J Phys Anthropol 2004;123:119–129.

John R. Lukacs, PhD
Department of Anthropology
University of Oregon
Eugene OR 97403 (USA)
Tel. +1 541 346 5112, Fax +1 541 346 0668, E-Mail jrlukacs@uoregon.edu

Koppe T, Meyer G, Alt KW (eds): Comparative Dental Morphology.
Front Oral Biol. Basel, Karger, 2009, vol 13, pp 162–166

Can Dental Caries Be Interpreted as Evidence of Farming? The Asian Experience

N. Tayles[a] · K. Domett[b] · S. Halcrow[a]

[a]Department of Anatomy and Structural Biology, University of Otago, Dunedin, New Zealand; [b]School of Medicine and Dentistry, James Cook University, Townsville, Australia

Abstract

The seminal development of control of food production and its social and biological effects on human populations has for a long time been one of the foci of prehistoric research. The relationship between diet and oral pathology is well recognised and accepted to the point where rates of dental caries in particular have been seen as indicative of subsistence mode. This is despite the complex aetiology of caries, with both genetic and environmental factors other than diet contributing to lesion frequency. Most publications considering prehistoric diet and caries acknowledge the contribution of non-dietary variables but provide a more comfortable dietary explanation, with the role of domesticated starchy staples paramount. This widespread acceptance of a simple relationship between dental pathology and starchy carbohydrates needs to be challenged, as there is no reason why one dietary component would be solely responsible for the development of caries or why all carbohydrates should have the same effect. Some years ago, on the basis of evidence from prehistoric rice farming communities in Southeast Asia, we questioned the relationship between dental caries and the presumptive increased carbohydrate consumption consequent to the adoption of agriculture. This paper reviews recent literature on the topic and presents evidence that there is still no simple or universally applicable explanation for patterns of changes in caries frequencies during human prehistory.

The control of food production afforded by the introduction of agriculture in late prehistory is seen as having had profound demographic, social and political consequences for human populations [1–3] but how was quality of life affected? Was human health better or did more people survive but endure a less than ideal existence?

In relation to oral health, the increase in agriculturally produced carbohydrates has been seen as a driver for a rapid increase in one of the frequently cited health indicators, that of dental caries, along with other aspects of oral health such as infections (i.e. dental abscesses) and tooth loss [4–6]. A publication by Christy Turner II in 1979 [7] had indeed set the scene for this concept by presenting data on dental caries from multiple collections of prehistoric human skeletal remains. These collections were classified as having come from one of three economies 'hunter-gatherer', 'agricultural' or euphemistically 'mixed'. Using this classification, Turner presented an apparently well defined, worldwide, dichotomy in caries frequency between non-agriculturalists, with low rates, and agriculturalists, with high rates. Euphemistically 'mixed' economies conveniently fell between the two extremes.

This paper presented such a splendidly simple story, appealing to the human desire to understand complexity by categorising, that it beguiled, and continues to beguile, many who read it so that it continues to be cited as authoritative almost 30 years after publication. This acceptance has been so complete that caries rates in numerous skeletal collections have been used as the basis for interpretation of the economies of the societies represented as agricultural, forager/hunter-gatherer, or 'mixed' [8, 9]. The concept that caries rates correlate perfectly with farming is very appealing but unfortunately biology (and human biology in particular) and subsistence modes are never that simple.

In 2000, we [1] published a paper considering the range of caries rates in three late prehistoric Southeast Asian sites. These sites covered the period from early rice availability to intensive production. On the basis of the rates at these sites, we suggested that the concept of increased caries rates with agricultural intensification (we did not have any 'pre-agricultural' sites to allow us to comment on agricultural origins) did not apply in prehistoric Southeast Asia, and that there is no clear evidence that rice farming, and the associated increase in carbohydrates in the diet, would necessarily lead to increased caries.

In the years since 2000 it has become clear that there is a lot more complexity in all aspects of the topic that need to be considered. Since the development of dental caries requires dietary carbohydrates for appropriate bacteria in dental plaque to ferment, creating acidic by-products that have the ability to demineralise dental tissues, clearly diet is a contributor to rates of the disease. Recently, significant advances have been made in clinical understanding of the complexity of caries aetiology and epidemiology, showing that rates can never be attributed to a single variable. Caries is a complex, multifactorial disease resulting from an interaction of numerous intrinsic and extrinsic factors, such as saliva flow rate and pH and dental morphology, and environmental factors such as exposure to fluoride, along with 'risk-conferring behaviours' [10]. The risky behaviour includes, but is not limited to, the consumption of cariogenic carbohydrates. Frequency of eating and the consequent maintenance of low oral pH is another contributor, and regardless of diet, caries will not develop where an effective level of oral hygiene is maintained [11].

In addition to all the non-dietary variables that can affect caries rates is the issue that not all carbohydrates are created equal. The carbohydrate involved, whether a mono- or disaccharide sugar or a complex starch, is most important. Sugars alone are highly cariogenic whereas starches are relatively mildly cariogenic, although starch consumed in combination with sugars increases cariogenicity [12]. There is considerable variability in the cariogenic potential of starches, related to the rates at which they can be cleared or fermented. This in turn partly reflects their structure, the method of processing, and the extent to which they are gelatinised during cooking. The more unrefined plant foods stimulate saliva flow, clearing the food debris from the mouth as well as buffering plaque acid and promoting remineralization of enamel [12]. Rice has been shown experimentally to have low intrinsic cariogenicity, particularly when it is consumed in its least processed form. This is best illustrated by sticky or glutinous rice (*Oryza sativa var. glutinosa* or *Oryza glutinosa*), frequently eaten in northeast Thailand today, which paradoxically does not adhere to the teeth because of the processing method [13].

Bearing these factors in mind, what has been said since our 2000 paper was published about the relationship of caries rates in prehistoric populations to the development of rice as an agricultural crop?

In 2006, Oxenham et al. [14] reviewed the evidence of 'oral health' (caries rates, infections/inflammations, and ante mortem tooth loss) from a series of prehistoric sites in Thailand and Vietnam, including our original three Thai sites, and concluded that there is indeed no evidence for a decline in oral health associated with the adoption and/or intensification of rice agriculture. They argue that an apparently later development of social complexity in Southeast Asia than elsewhere in the world correlates with the maintenance of production at the household level, with the consequence that large-scale agricultural production did not develop until after the advent of iron, delaying the onset of an increase in caries rates to a period beyond that reflected in the sites reviewed.

The other relevant research is that by Temple and Larsen [15] on foraging and rice-agricultural populations in Japan. Larsen has long been a proponent of the relationship between caries and agriculture, principally on the basis of the evidence from North American maize agriculture. In a co-authored paper [15] using samples of pre-agricultural Jomon and immigrant rice farmers, the Yayoi, the authors found that caries rates in only one of three Yayoi samples were significantly higher than those of Jomon. What is also interesting is the significant variation between the rate in one Yayoi group compared with the other two, although Temple and Larsen appear to have been pursuing an agenda, as by the end of their paper they are nevertheless claiming a 'generally greater frequency of carious teeth' [15, p. 507] among the Yayoi. Nevertheless, they ultimately suggest that the variation in frequencies among the Yayoi might reflect 'dietary choices' [15, p. 507] involving different foods. The combination of rice with other foods can certainly alter caries rates but we suggest that the variation is equally likely to reflect variation in the method of processing or cooking rice.

An example of a prehistoric population from Taiwan with rice agriculture, included by the authors in a 'mixed' subsistence mode with some contribution from 'cereal crops, such as rice' to hunting, fishing and collecting of marine resources is published by Pietrusewsky and Tsang [16, p. 206]. The small sample of 23 individuals had very low frequencies of dental caries, and of most other indicators of oral health. The authors use the caries rates to argue the group was 'non-agricultural or an economy that included a subsistence diet that was low in starches and sugars' [16]. Again, it is possible that the rice was consumed in a minimally processed form. The low caries rates do not warrant excluding agriculture as a contributor to diet.

To summarise this small group of publications, there is as yet no consensus on the relationship between caries and rice agriculture. There are several issues that need to be considered before such a consensus might be reached. Firstly, was the development of agriculture truly an instantaneous event, at least on an archaeological timescale? This is the traditional viewpoint, arising from the belief that it was a 'revolution'. Some still argue this to be the case [17], although the concept of an agricultural transition, a 'naturally occurring process of great selective value' [18, p. 13] seems a much more appropriate descriptor, allowing for variability in both speed and scale of economic change. Beyond the contribution to diet of a starchy 'staple' carbohydrate is the question of the range of diet, particularly the inclusion of sugars.

Secondly, the question of the variability of cariogenicity of rice, or any other starchy carbohydrate for that matter, cannot be ignored. The differential effects of the variety or species being consumed and the degree of processing have already been mentioned. In 2000, we argued that the level of processing of rice could well be a factor affecting its contribution to caries rates, as it is most likely that rice was eaten unpolished, with the coarse texture both reducing its stickiness and stimulating salivary flow, increasing clearance rates.

Thirdly, caries is now well-recognised to be a multifactorial disease. Diet can be a major contributor but it is not the only determinant of prevalence as noted above.

Last but not least is the perennial issue of interobserver variation in the recording of caries.

Weslowski [19] evaluated 26 papers on dental caries in prehistoric populations published between 1999 and 2004 and from this showed that a majority failed to detail their methods of caries diagnosis. This renders accurate evaluation of the comparability of prevalence data impossible. She also showed that very few of the papers gave adequate information on relevant characteristics of the sample such as age and sex, and most did not consider any contributor to prevalence other than diet.

As many researchers in the field of bioarchaeology come from the social rather than the biological sciences, the tendency to favour a behavioural explanation is understandable, but not excusable. While recognizing that many of the factors contributing to caries aetiology cannot be known for skeletal samples, to continue to advocate the concept that caries rates can be used as a proxy for a subsistence mode, agricultural or otherwise, is naïve.

The recent excavation of a significant Thai site, Ban Non Wat, has provided a large sample of 637 burials dating from 1700 BC to 400 AD. The sample represents the community from pioneer settlers in a small village to inhabitants of a substantial, sophisticated settlement with strong evidence of hierarchy/inequality. This is a period of increasing social complexity but without clear evidence of agricultural surpluses, arguing against the universal coupling of socio-political changes with agricultural intensification. The analysis of the evidence from these burials is work in progress at the time of writing but for the first time there is a possibility of addressing the question of the role of intensification of rice agriculture on dental caries frequencies and on oral health in general in a large sample from a single site.

References

1 Tayles N, Domett K, Nelsen K: Agriculture and dental caries? The case of rice in prehistoric Southeast Asia. World Archaeol 2000;32:68–83.

2 Diamond J, Bellwood P: Farmers and their languages: the first expansions. Science 2003;300:597–603.

3 Harris DR: The Origins and Spread of Agriculture and Pastoralism in Eurasia. London, UCL Press, 1996.

4 Larsen CS: Biological changes in human populations with agriculture. Ann Rev Anthropol 1995;24:185–213.

5 Arnay-de-la-Rosa M, González-Reimers E, Gámez-Mendoza A, Galindo-Martín L: The Ba/Sr ratio, carious lesions, and dental calculus among the population buried in the church La Concepción (Tenerife, Canary Islands). J Archaeol Sci 2008;36:351–358.

6 Keenleyside A: Dental pathology and diet at Apollonia, a Greek colony on the Black Sea. Int J Osteo 2008;18:262–279.

7 Turner CG II: Dental anthropological indications of agriculture among the Jomon people of Central Japan. Am J Phys Anthropol 1979;51:619–636.

8 Bernal V, Novellino P, Gonzalez PN, Perez SI: Role of wild plant foods among Late Holocene Hunter-Gatherers from Central and North Patagonia (South America): an approach from dental evidence. Am J Phys Anthropol 2007;133:1047–1059.

9 Delgado-Darias T, Velasco Vázquez J, Arnay-de-la-Rosa M, Martín-Rodríguez E, González-Reimer E: Dental caries among the Prehispanic population from Gran Canaria. Am J Phys Anthropol 2005;128:560–568.

10 Fejerskov O: Changing paradigms in concepts on dental caries: consequences for oral health care. Caries Res 2004;38:182–194.

11 Nyad B: The role of oral hygiene; in Fejerskov O, Kidd E (eds): Dental Caries: The Disease and Its Clinical Management. Oxford, Blackwell, 2004, pp 172–177.

12 Moynihan P, Lingström P, Rugg-Gunn AJ, Birkhed D: The role of dietary control; in Fejerskov O, Kidd E (eds): Dental Caries: The Disease and Its Clinical Management. Oxford, Blackwell, 2004, pp 223–244.

13 Techanitiswad T: Determinants of the variation in dental caries of 2- and 4-year-old Thai children in an urban area of Khon Kaen Province, North Eastern Thailand; unpubl PhD thesis, University of Otago, Dunedin, 1994.

14 Oxenham M, Nguyen L, Nguyen K: The oral health consequences of the adoption and intensification of agriculture in Southeast Asia; in Oxenham M, Tayles N (eds): Bioarchaeology of Southeast Asia. Cambridge, Cambridge University Press, 2006, pp 263–289.

15 Temple DH, Larsen CS: Dental caries prevalence as evidence for agriculture and subsistence variation during the Yayoi Period in Prehistoric Japan: biocultural interpretations of an economy in transition. Am J Phys Anthropol 2007;134:501–512.

16 Pietrusewsky M, Tsang C-H: A preliminary assessment of health and disease in human skeletal remains from Shi San Hang: a prehistoric aboriginal site on Taiwan. Anth Sci 2003;111:203–223.
17 Bellwood P: First Farmers: The Origins of Agricultural Societies. Malden, Blackwell, 2005.
18 Higgs ES, Jarman MR: The origins of animal and plant husbandry; in Higgs ES (ed): Papers in Economic Prehistory. Cambridge, Cambridge University Press, 1972, pp 3–13.
19 Weslowski V: Caries prevalence in skeletal series: it is possible to compare? Mem Inst Oswaldo Cruz 2006;101:139–145.

Dr. Nancy Tayles
Department of Anatomy and Structural Biology, Otago School of Medical Sciences, University of Otago
PO Box 913, Room 331, Lindo Ferguson Building
Dunedin (New Zealand)
Tel. +64 3 479 7372, Fax +64 3 479 7254, E-Mail nancy.tayles@anatomy.otago.ac.nz

Koppe T, Meyer G, Alt KW (eds): Comparative Dental Morphology.
Front Oral Biol. Basel, Karger, 2009, vol 13, pp 167–172

Teeth and the Past in Portugal: Pathology and the Mesolithic-Neolithic Transition

Mary Jackes

Department of Anthropology, University of Waterloo, Waterloo, Ont., Canada

Abstract

Carious lesions are considered an important marker of dietary change at the transition from hunting and gathering to horticulture. Within the context of the transition to the Neolithic in Central Portugal, this paper discusses factors which must be taken into consideration in reporting dental pathology frequencies. Three sites are examined, two late Mesolithic shell middens and one early Neolithic burial cave dating before 5500 calBP which is taken to be the end of the transition period. Comparability of results across different burial types and depositional environments requires close attention to methodology. Despite inclusion of necessary detail on caries type, age of onset of pathology and age distribution within the sample, factors such as the use of teeth as tools and post-mortem alteration of teeth may make it impossible to be certain of rates of pathology. Inter-site differences in dental pathology may result, not only from diet, but from differing adult age distributions: when burial modes and deposits are dissimilar, differing diagenesis and taphonomy may further bias pathology rates, as well the use of teeth as tools which can affect attrition, trauma and tooth loss rates.

Cook [1] suggests that a caries rate change is the only reliable marker of the transition to agriculture. The transition period is particularly well represented by skeletal material in Portugal at the Mesolithic shell middens, Moita do Sebastião (Moita) and Cabeço da Arruda (Arruda), dating after 8000 calBP and located on a southern tributary of the Tagus River and at the Early Neolithic cave site of Casa da Moura, north of the Tagus. Arruda dates extend slightly later in time than those for Moita [2] and the transition is defined as prior to 5500 calBP. The transition for these sites is clearly marked by differences in stable isotope values and in burial modes [3].

The primary burials at Moita and Arruda were mixed before and after excavation: the minimum number of individuals (MNI) and their age distribution is best assessed by seriation of the mandibles [4] providing a 'maximum' MNI (Moita, 85; Arruda, 105 [2]). The Casa da Moura skeletons are disarticulated and 74% of the teeth were loose when excavated. Reference to teeth in situ in their alveoli and multiple sorting with varying emphases on crown features, root form and metrical analyses of size and shape has aided identification of these loose teeth. The mandibles with teeth still in the alveoli could be seriated and were analyzed in Portugal in the same way as the Mesolithic mandibles, but analyses of the nearly 5,000 loose teeth were carried out in a laboratory in Canada over several years and included inter- and intra-observer error tests. The MNI for Casa

Table 1. Percentage rate of dental pathology within each of ten (A–J) more or less equal groupings of molar sockets across seriated mandibles

	Moita do Sebastião			Cabeço da Arruda		
	caries	abscessing	pre-mortem tooth loss	caries	abscessing	pre-mortem tooth loss
A	16.0	0.0	0.0	0.0	0.0	0.0
B	0.0	0.0	0.0	0.0	0.0	0.0
C	20.0	0.0	0.0	3.8	0.0	0.0
D	20.8	3.6	3.6	3.7	0.0	0.0
E	10.0	0.0	4.2	18.5	0.0	12.5
F	0.0	0.0	11.5	19.2	0.0	0.0
G	16.0	3.6	3.6	8.3	3.2	16.1
H	19.0	3.6	21.4	12.9	3.0	8.3
I	29.4	24.0	20.0	20.0	8.6	28.2
J	50.0	15.8	78.9	18.8	14.3	41.7
Present/n	31/211	12/256	32/256	25/249	10/309	38/323

da Moura has been estimated as 340 based on M_2 intact teeth and sockets with pre-mortem tooth loss [2]. No doubt some older individuals must be unrepresented in the count.

In considering the value of dental pathology as central evidence for subsistence change, I examine lower molars because good preservation is a characteristic of mandibles and particularly of lower molars. Caries susceptibility depends upon tooth type [5], and tooth type representation in archaeological material is very uneven. Lower molars are best represented [6, 7]: limiting analysis to lower molars ensures comparability between Mesolithic primary and Neolithic ossuary samples.

Periodontal disease will not be discussed here partly because of the presence of matrix or damage. The height of the cemento-enamel junction (CEJ) above the alveolar margin has been recorded for all sites but cannot be discussed in this brief summary.

The Mesolithic mandibles assessed at age ~15 years and over were seriated on attrition. Attrition rate differences, even between the Mesolithic sites, are likely [8]. Moita individuals of ~20 years of age have a higher level of wear for M_1 relative to M_2 and M_3 than Arruda individuals of equivalent age. Sorting by wear stages suggests that Arruda has more older people than Moita: based on hazard rate analyses, significantly more adults survived into higher wear level stages at Arruda than at Moita. This is supported by analyses of Nordin's Index and cortical width and density, all features of the femoral cortex [9].

In table 1 the lower molar sockets for the Mesolithic sites are distributed over ten categories (A–J) with near equal numbers of sockets, in an attempt to understand the relationship of dental pathology to increasing age. This technique of distributing individuals is designed to decrease the effects of unequal wear stage samples and any intra-observer change in wear assessments between the two sites. Moita pathology occurs earlier, despite the possibility of faster wear. This result is equivalent to previous 11 attrition grade distribution results: caries is seen across wear stages at Moita, but is rare in younger individuals at Arruda [9].

At Moita, carious lesions occur before M_3 comes into wear and multiple lesions in one individual are seen as soon as M_3 reaches the first stage of wear. Multiple lesions are not seen at Arruda until category E in an unusual individual with four caries in association with an M_1 having only a trace of mid-occlusal surface enamel left. Tooth loss also occurs earlier at Moita than at Arruda. Abscessing and premortem tooth loss are not seen at Arruda until the second molars have large discrete dentin exposures, whereas they occur when Moita second molars still have no more than point exposures.

Moita caries and tooth loss cannot be ascribed to heavy attrition alone. Moita occlusal lesions occur in the form of small pit and fissure defects in four individuals (in categories A–G) with M_2 enamel barely worn to pinpoint dentin exposure. Open pulp chambers are not seen at Moita (in two individuals in category I) until M_2 occlusal enamel survives only as a narrow marginal rim. All carious individuals in the midrange of attrition have multiple lesions on both occlusal and interproximal surfaces.

Stable isotope values differ significantly between the two sites [2, 3], although the distributions overlap. This suggests varying degrees of reliance on estuarine resources and some of the heterogeneity of Mesolithic stable isotope values may reflect changes within the estuarine setting which led eventually to abandonment of the valley [2].

Casa da Moura pathology rates are lower than for the Mesolithic sites [3], but pathology is found in young individuals, even in deciduous dentitions, while no Mesolithic deciduous teeth had lesions. Stable isotopes indicate a homogeneous Early Neolithic diet from terrestrial resources [3].

An initial sample of the in situ Casa da Moura dentitions was observed in 1986, but most were examined in 1989 after detailed study of the loose material. Further examination is ongoing. Overall, in those over the age of ~15 years in the 1989 sample, there was a 9.9% caries rate (27/273) and a 10.5% premortem tooth loss rate (32/305). Equivalent figures for the Mesolithic sites (table 1) are Moita (12.5 and 14.7%) and Arruda (10.0 and 11.8%).

By examining only those mandibles in the 1989 sample with unexpanded points of dentin exposure on M_1, we see that caries occurs early at Casa da Moura. Of the molars in occlusion (103), 12 had caries: two had occlusal pits, two had interproximal lesions just below the facet, two had the interproximal defect enlarged to involve the occlusal surface, two defects at the CEJ in one individual, one carious buccal pit, and one large occlusal cavity. One further individual had two defects, one beginning at the mesial CEJ and the other a huge multi-surface lesion.

Caries type changes across time [3, 6] with occlusal caries rates decreasing between Moita and Arruda ($p = 0.019$) and again between Arruda and Casa da Moura (in situ 1986–89 molars $p = 0.03$, loose $p = 0.015$), while interproximal caries incidences show no significant differences across sites.

The co-occurrence of strong molar attrition and occlusal caries at Moita [10] and the reduction in dental pathology in the late Mesolithic and Early Neolithic perturb the expected pattern [11, 12]. In central Portugal, markedly lower attrition levels (fig. 1a) and a rise in occlusal caries rates do not occur until later in the Neolithic, after 5500 calBP.

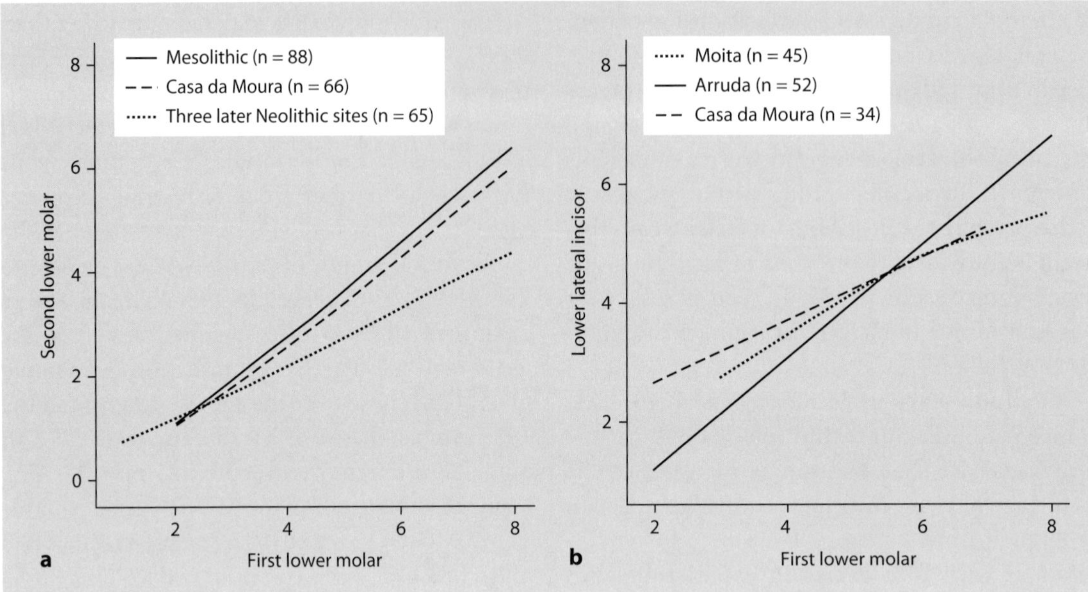

Fig. 1. Loess fit lines using the Epanechnikov kernel to smooth 99% of data points. **a** Comparison of the relative attrition of the permanent first and second lower molars in Portuguese Mesolithic and Neolithic individuals indicates the major change in attrition of lower molars occurred after 5500 calBP. **b** Comparison of the relative attrition of the permanent lower lateral incisors and first lower molars in Portuguese Mesolithic and Neolithic individuals indicates that Casa da Moura had earlier, stronger anterior wear than either of the Mesolithic sites.

Nevertheless, Casa da Moura attrition is reduced. To characterize Early Neolithic loose molar wear adequately, a composite of the approaches of Smith [13] and Lovejoy [14] was developed. Of eight crown heights, the disto-buccal is most closely related to this composite scheme (fig. 2) and this variable is used in comparisons across the 937 Casa da Moura loose lower molars.

A large sample (586) of the loose first and second lower molars was reexamined for caries to record inter-observer differences. The only disagreements among observations occurred with interproximal changes at the CEJ, coded either as grooving or as caries [6]. (Note that the grooves discussed here did not result from the use of toothpicks. Toothpick grooves occur and are easily distinguished from CEJ grooving [6].)

To What Extent Is Grooving Carious or Post-Mortem?

Grooving at the CEJ was studied in detail in Casa da Moura loose canines, in which it occurs without relation to hypercementosis, rotation, crown dimensions, attrition, or jaw. A further consideration is the use of anterior teeth in leatherworking: identified as a factor in Mesolithic anterior dental attrition [15], the Early Neolithic anterior wear is more extreme. Examination at up to ×1,000 shows that consistently oriented criss-crossing striations resulted from this apparent use of the anterior teeth. Such lingual polishing occurs mostly in the maxilla and most strongly on the left (left upper canines show lingual polish significantly more often than the right). Casa da Moura mandibular anterior attrition

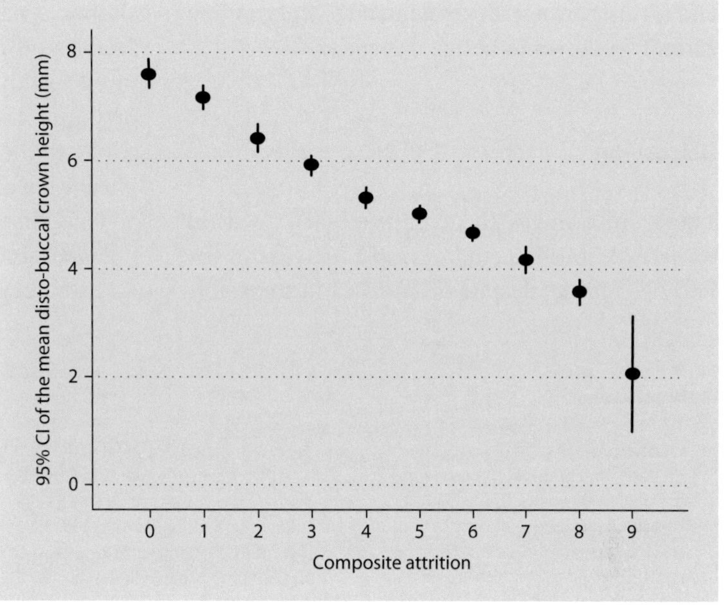

Fig. 2. Casa da Moura composite attrition and the disto-buccal crown height (323 lower molars): loess line at 99% (Epanechnikov) is nearly identical with a lineal regression line, r = 0.840.

is established earlier and more strongly than at Moita and Arruda (fig. 1b): lateral incisors are graphed because the central lower incisors are lost pre-mortem significantly more frequently than the laterals. It is clear that Casa da Moura teeth are affected by their use as tools. Grooving on the anterior tooth crowns, as distinct from the CEJ, is also attributable to leatherworking. There is no relationship between CEJ grooving and canine lingual polish.

Trauma of the canines is also common. Trauma in the lower canines occurs significantly more often on the left (of which 26% are chipped) than on the right ($p = 0.004$), and major breakage occurs only on the left. However, CEJ grooving is not related to the presence, type or location of trauma. CEJ grooves are unlikely to be the result of the type of trauma that gives rise to abfraction [16].

Much CEJ grooving seems to be post-mortem [6]. Grooving is unlikely to significantly alter caries incidence in canines, but it will certainly affect the loose Casa da Moura lower molar caries rate. Of the interproximal lesions for which the origin could be pinpointed, 11 were at the CEJ, and only three were subfacet. CEJ grooving cannot always be distinguished from caries with complete certainty. The grooving might also be initiated by carious lesions and post-mortem grooving could mask the presence of such lesions. In sites in which this type of diagenesis is common, the caries rate is uncertain.

CEJ grooving cannot be clearly identified as specific to individuals at higher wear levels (reduced disto-buccal crown heights). In a sample of 221 lower molars in which attrition was at least at the level of two or three pinpoint exposures of dentin, 17 had CEJ grooving. Neither crown height means nor variances were significantly different between those with and without grooving: the difference in means between carious molars (28) and those without caries (192) was significant ($p = 0.04$) and trauma is also significantly associated with lowered disto-buccal crown height ($p = 0.000$). There is no association between CEJ grooving and enamel chipping ($p = 0.5$). Caries and trauma are age-dependent, but CEJ grooving cannot be shown to be age-dependent.

Discussion

There are changes in dental pathology within the late Mesolithic, but no simple increase in the Early Neolithic, despite differences in the stable isotopes. Analysis by wear levels and by metric variables systematically altered by attrition is required to fully describe the changes. Simple reporting of caries rates, without attention to the age bias in the sample, without examining the age distribution of pathology, without recording type and location of lesions and without consideration of diagenesis and tooth type, is not likely to give us a fully accurate picture of the past.

References

1 Cook DC: Maize and Mississippians in the American midwest 20 years later; in Cohen, MN, Crane-Kramer GMM, Larsen CS (eds): Ancient Health: Skeletal Indicators of Agricultural and Economic Intensification. Bioarchaeological Interpretations of the Human Past: Local Regional and Global Perspectives. Gainesville, University Press of Florida, 2007, pp 10–19.

2 Jackes M, Meiklejohn C: The Paleodemography of Central Portugal and the Mesolithic-Neolithic transition; in Bocquet-Appel JP (ed): Recent Advances in Paleodemography: Data, Techniques, Patterns. Dordrecht, Springer, 2008, pp 209–258.

3 Lubell D, Jackes M, Schwarcz H, Knyf M, Meiklejohn C: The Mesolithic-Neolithic transition in Portugal: isotopic and dental evidence of diet. J Archaeol Sci 1994;21:201–216.

4 Jackes M, Meiklejohn C: Building a method for the study of the Mesolithic-Neolithic transition in Portugal; in Budja M (ed): Neolithic Studies 11. Documenta Praehistorica 2004, vol 31, pp 89–111.

5 Batchelor PA, Sheiham A: Grouping of tooth surfaces by susceptibility to caries: a study in 5–16 year-old children. BMC Oral Health 2004;4:2. http://www.biomedcentral.com/1472–6831/4/2

6 Jackes M, Lubell D: Dental pathology and diet: second thoughts; in Otte M (ed): Nature et Culture: Actes du Colloque International de Liège, 13–17 Décembre 1993. Liège, Études et Recherches Archéologiques de L'Université de Liège, 1996, no 68, pp 457–480.

7 Jackes M: Representativeness and Bias in Archaeological Skeletal Samples; in Agarwal S, Glencross B (eds): A Handbook of Social Bioarchaeology. Blackwell Studies in Global Archaeology. Oxford, Wiley-Blackwell, in press.

8 Jackes M, Lubell D: Human skeletal biology and the Mesolithic-Neolithic transition in Portugal; in A. Thévenin (ed.), dir. scientifique P. Bintz: Europe des derniers chasseurs Épipaléolithique et Mésolithique: actes du 5e colloque international UISPP, commission XII, Grenoble, 18–23 septembre 1995. Paris, Éditions du CTHS, 1999, pp 59–64.

9 Lubell D, Jackes M: Portuguese Mesolithic-Neolithic subsistence and settlement. Riv Antropol 1988;LXVI(suppl): 231–248.

10 Meiklejohn C, Wyman JM, Schentag CT: Caries and attrition: dependent or independent variables? Int J Anthropol 1992;7:17–22.

11 Meiklejohn C, Zvelebil M: 1991 Health status of European populations at the agricultural transition and implications for the causes and mechanisms of the adoption of farming; in Bush H, Zvelebil M (eds): Health in Past Societies. Br Archaeol Rep Int Ser 567:129–146.

12 Jackes M, Lubell D, Meiklejohn C: Healthy but mortal: human biology and the first farmers of Western Europe. Antiquity 1997;71:639–658.

13 Smith BH: Patterns of molar wear in hunter-gatherers and agriculturalists Am J Phys Anthropol 1984;63:39–56.

14 Lovejoy CO: Dental wear in the Libben population: its functional pattern and role in the determination of adult skeletal age at death. Am J Phys Anthropol 1985;68:47–56.

15 Lefèvre J: Étude odontologique des hommes de Muge. Bull Mem Soc Anthropol Paris 1973;10:301–333.

16 Palamara JEA, Palamara D, Messer HH, Tyas MJ: Tooth morphology and characteristics of non-carious cervical lesions. J Dent 2006;34:185–194.

Dr. Mary Jackes
Department of Anthropology, University of Waterloo,
Waterloo, ON N2L 3G1 (Canada)
Tel. +1 519 888 4567 ext. 32520, Fax +1 519 747 9149, E-Mail mkjackes@uwaterloo.ca

Koppe T, Meyer G, Alt KW (eds): Comparative Dental Morphology.
Front Oral Biol. Basel, Karger, 2009, vol 13, pp 173–177

Pattern of Dental Caries in the Historical Human Population of Kujawy in the Polish Lowland (North-Central Poland)

Marta Kurek · Beata Borowska-Strugińska · Julia Mazurkiewicz · Joanna Nieczuja-Dwojacka · Elżbieta Żądzińska

Department of Anthropology, University of Łódź, Łódź, Poland

Abstract

The aim of this study was to determine the dental caries rates in the skeletal human population who lived in north-central Poland in modern times (16th–18th century). The material consisted of 297 skeletons (4,783 permanent teeth) from the archeological site of K. Compared material was coming from two different sites located in the same region: early mediaeval Kolonia – 39 skeletons and Święty Duch (dating to the same historical period as K population) – 417 skeletons. All three series are part of the osteological collection of the Department of Anthropology. For each skeletal sample, caries rates were calculated by individual (frequency index) and by tooth count (intensity index). Dental caries was scored also according to a procedure in which the initiation sites are recorded. The highest frequency of dental caries was observed in Święty Duch (70%) and the lowest in the early mediaeval sample (almost 59%). The intensity of caries fluctuated from above 8% observed in Kolonia to 22% observed in the sample from Święty Duch. However, in all three samples the caries initiation site was most often noticed on fissure and pits (type 1) the K population was characterised by a different caries pattern ($\chi^2 = 17.88$, $p < 0.05$ when compared with that of Kolonia; $\chi^2 = 218.73$, $p < 0.01$ when compared with that of Święty Duch), which could be the effect of diet (high frequency of CEJ caries probably means starch-rich plant food diet), oral hygiene, fluoride level, economic status or genetic factors (resistance/sensitivity to cariogenic bacteria).

The pathological processes observed in the human stomatognathic system are a good gauge of the general health of a population. A disease that often occurs in the teeth is dental caries, an analysis of which makes it possible to assess the health, hygiene and diet of the human population under consideration.

Dental caries is defined as an organic pathological process caused by extrasomatic factors. It is connected with demineralisation and subsequent decomposition of dental hard tissues: enamel, dentine, and cement [1].

The development of carious lesions is conditioned by the concurrence of five etiological factors, including: genetically determined dental susceptibility (the occurrence of alleles related to resistance/vulnerability to carious processes) [2], the presence of acidogenic microorganisms in the oral cavity (e.g. *Streptococcus mutans*,

Streptococcus salivarius, Lactobacillus acidophilus, Actinomyces, and *Neisseria),* the presence of carbohydrates in human diet, the degree of hardness of tooth tissues compared with contents of inorganic matter (for example fluoride level), tooth morphology (natural furrows and pits in the crown) as well as the duration and frequency of exposition of dental tissues to pathogenic factors, oral hygiene and diet [3].

The objective of the study was to assess the dental caries rates in the human population inhabiting Kujawy (the central part of the North European Lowland) at the turn of the Middle Ages and the modern era (16th–18th centuries).

Materials and Methods

Dental caries was assessed in the population from site K, dating to the 16th–18th centuries, in Brześć Kujawski the main city of the Kujawy (north-central Poland, the area of the North European Lowland, which is documented to have been continuously inhabited by human populations since the Early Neolithic) [4].

The sample under consideration contained 504 skeletons, of which 297 skulls (4,783 teeth) were used in this study, 41.41% of them being subadults skulls.

Comparative material included skeletal series from archaeological sites located in the same town: from the Święty Duch site – burial mound dating to the same historical period but located separately (4,320 teeth from 417 individuals, 13% subadults) and from the early mediaeval (12th century) sites in Kolonia (679 teeth from 39 individuals, 20.5% subadults) [5].

The examination of the dental remains consisted in an evaluation of the presence of caries in all types of teeth in accordance with the methodology proposed by Moore and Corbett [6, 7], classifying lesions on a scale of six classes, taking into consideration the initiation site of the lesion: 1 = fissure or pit, 2 = approximal surface, just below contact point, 3 = cervical surface, 4 = cement-enamel junction, 5 = root surface, 6 = gross (when the crown is so destroyed that no site of initiation can be deduced).

The dental caries frequency – prevalence by individual (the percentage of individuals affected by caries in the total number of individuals under investigation) as well as the caries intensity rate – prevalence by tooth count (the percentage of teeth with carious lesions in the total number of teeth under investigation) [1, 7] were calculated. Additionally the caries correction formula proposed for skeletal series by Lukacs [8] was used, where the number of teeth lost antemortem were multiplied by the percentage of teeth with caries-induced pulp exposure to produce an estimate of the number of teeth lost antemortem due to caries. This value was then converted into the corrected dental caries rate by adding the number of teeth observed to have carious lesions and the estimated number of teeth lost antemortem because of caries and then dividing this number by the total number of teeth observed plus all teeth known to have been lost antemortem.

The χ^2 test was used to assess differences in the distribution of caries rates (corrected dental caries rate) as well as caries type within and between sites (by sex and by age).

Results

The frequency of caries in site K was 66.7%, while the intensity rate of dental caries was 14.8%.

Although the χ^2 test for independence did not reveal a difference in the frequency and the intensity of the disease between the sexes ($\chi^2 = 0.86$, $p > 0.05$), the patterning of caries noticed according to the 6-grade scale was distinctly dimorphic ($\chi^2 = 14.34$, $p < 0.05$). Caries on female teeth in comparison to male teeth is more often initiated on the approximal surface, just below the contact point. The χ^2 test did not reveal the age differences in caries rates ($\chi^2 = 0.20$, $p > 0.05$ for prevalence by tooth, $\chi^2 = 7.14$, $p > 0.05$ for initiation site rate) (tables 1, 2). Carious cavities were equally often observed in both jaws; in the maxilla carious teeth amounted to 15.8%, and in the mandible 13.3% ($\chi^2 = 2.45$, $p > 0.05$).

Type 1 predominates in carious teeth, where the primary site of the pathological process is the masticatory surface (molars), followed by type 6, where due to the large scale of the cavity, the primary site cannot be identified. The least frequent was type 5 – a lesion initiated on the root surface (under the CEJ).

These results were compared with the previously examined series (fig. 1). The population from site K differed in terms of the caries 'pattern'

Table 1. Dental caries rates by tooth count

Site	Male				Female				Adult sexes pooled			
	observed		corrected		observed		corrected		observed		corrected	
	%	n/total	%	n/total	%	n/total	%	n/total	%	n/total	%	n/total
K	14.3	241/1,682	14.8	253/1,712	15.2	296/1,949	15.9	317/1,996	14.8	537/3,631	15.4	570/3,708
Św. Duch	19.8	401/2,027	22.1	450/2,038	25.2	393/1,558	26.6	418/1,572	22.1	794/3,585	24.0	868/3,610
Kolonia	7.8	24/312	7.9	25/312	8.1	17/208	8.3	18/211	7.9	41/520	8.2	43/523

	Adult				Subadult			
	observed		corrected		observed		corrected	
K	14.8	537/3,631	15.4	570/3,708	14.8	170/1,152	14.8	171/1,153
Św. Duch	22.1	794/3,585	24.0	868/3,610	20.8	153/735	21.1	155/737
Kolonia	7.9	41/520	8.2	43/523	7.7	11/159	7.7	11/159

Table 2. χ^2 values for caries rate comparisons (calculated for corrected caries rate and caries initiation site rate) within populations and for sexes pooled between populations

Site	Adult vs. subadult		Male vs. female	
	corrected	initiation site	corrected	initiation site
K	0.20	7.14	0.86	14.34*
Św. Duch	3.09	15.56*	9.88**	21.13**
Kolonia	0.28	8.11	0.04	6.14
	Sexes pooled corrected		Sexes pooled initiation site	
K vs. Sw. Duch	87.13**		218.73**	
K vs. Kolonia	18.92**		17.88*	
Sw. Duch vs. Kolonia	66.56**		198.14**	

* p < 0.05, ** p < 0.01.

both from its contemporary population from the Święty Duch site ($\chi^2 = 218.73$, $p < 0.01$) and from the population inhabiting Kujawy in the early Middle Ages – the figures for the Kolonia site were $\chi^2 = 17.88$, $p < 0.05$.

Discussion

Caries is a multi-factor disease. A poor economic situation of populations, which can be accompanied by malnutrition, inferior food quality or

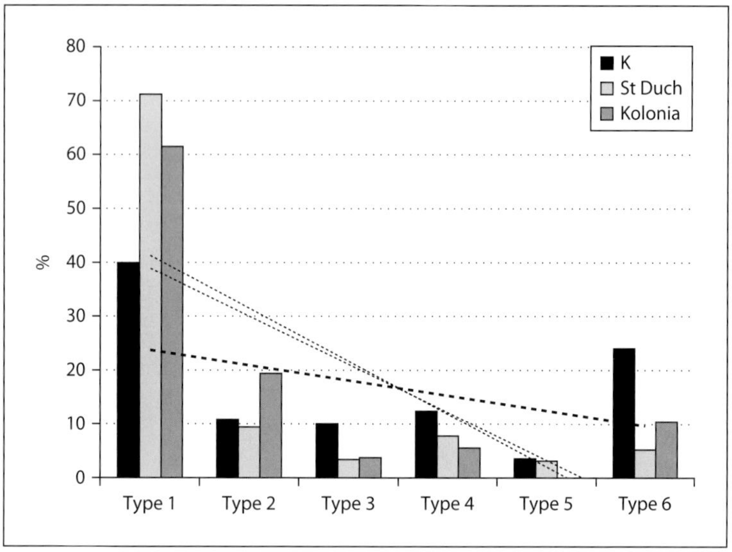

Fig. 1. Patterning of dental caries (with trend lines) in analysed human populations.

poor hygiene of the oral cavity, has an influence on dental caries morbidity.

The anthropological analysis of the K population carried out by Borowska-Strugińska [9] revealed that the population may have been characterised by a poor biological condition as compared with other Polish populations from that period, which was manifested in a different mortality distribution – higher mortality of children, a higher incidence of systemic diseases and a greater number of individuals with *cribra orbitalia* – a feature used as an indicator of biological stress.

In most populations, also in the population from the site Św. Duch, there are marked differences in the frequency and intensity of dental caries between the sexes. Females are characterised by a greater number of carious cavities, which is not only related to their social status and other environmental factors such as dietary preferences, but also due to the changes that affect a woman's body during pregnancy: hormonal changes and more frequent meals [10]. In the case of site K, sex differences in frequency and intensity of caries were not observed, which might be interpreted as strong dietary egalitarianism, confirmed by the contents of micro- and macroelements (Sr, Zn and Ca) in the teeth from other Polish archaeological sites both from the Neolithic and the Middle Ages [11]. On the other hand, sex dimorphism was noticed in the patterning of initiation sites of dental caries, which might confirm the 'hormonal' hypothesis proposed by Lukacs and Largaespada [10]. Sex differences in diet (also the higher frequency of starch-rich sneaking) as well as sex-dimorphic physiological factors (increase in cariogenic microbes in saliva during pregnancy) and specially the earlier eruption of female teeth could be responsible for the higher frequency of caries initiated on the approximal surface of teeth. Detailed studies of this disease (based on the Moore and Corbett scale) are very rare on Polish skeletal series, so it is likely that the sex differences in caries patterning have not been captured yet.

The large proportion of teeth with type 3 caries, where the initiation site was the surface of the crown just above the gingiva, and type 4 caries, where enamel demineralisation was initiated at the CEJ (as the result of lowering of the alveolar process) may indicate that the population had more unprocessed (coarse-grained) food stuffs and the starch-rich plant food diet, which would reflect its lower social status at that historical period [12].

Conclusion

The population from site K is characterised by a different caries rate and a different distribution of carious lesions as compared with other populations under consideration from the same area. This may indicate a different kind of susceptibility to caries, differences in social status, diet and oral hygiene, which might suggest a genetic or cultural dissimilarity of this population what could be helpful in understanding the population process towards the tumultuous end of the 17th century in Europe after the Swedish invasion of Poland (1655–1660) called in Polish history the Swedish Flood [13].

References

1 Alt KN, Rösing FW, Teschler-Nicola M: Dental Anthropology – Fundamentals, Limits and Prospects. Wien-New York, Springer, 1998.
2 Vieira AR, Marazita ML, Goldstein-McHenry T: Genome-wide scan finds suggestive caries loci. J Dent Res 2008;87:435–439.
3 Navia JM: Carbohydrates and dental health. Am J Clin Nutr 1994;59:719–727.
4 Grygiel R, Bogucki P: Early farmers in North-Central Europe: 1989–1994 excavations at Oslonki, Poland. J Field Arch 1997;24:161–178.
5 Żądzińska E: Dental caries among Mediaeval population from Stary Brześć Kujawski (central Poland). Slov Antropol 2004;6:204–207.
6 Kerr NW, Bruce MF, Cross JF: Caries experience in Mediaeval Scots. Am J Phys Anthrop 1990;83:9–70.
7 Hillson S: Dental Anthropology. Cambridge, Cambridge University Press, 1996.
8 Lukacs JR: Sex differences in dental caries rates with the origin of agriculture in South Asia. Curr Anthropol 1996;37:147–153.
9 Borowska-Strugińska B: Charakterystyka antropologiczna wybranej populacji zamieszkującej tereny Brześcia Kujawskiego w okresie od XVI do XIX wieku; doctoral thesis, Łódź, 2004.
10 Lukacs JR, Largaespada LL: Explaining sex differences in dental caries prevalence: saliva, hormones, and 'life-history' etiologies. Am J Hum Biol 2006;18:540–555.
11 Szostek K, Głąb H, Lorkiewicz W, Grygiel R, Bogucki P: The diet and social paleostratigraphy of Neolithic agricultural population of the Lengyel culture from Osłonki (Poland). Anthrop Rev 2005;68:29–41.
12 Gleń E: Stan uzębienia mieszkańców Krakowa od IX do XVIII w. (próchnica zębów). Przegląd Antropol 1975;41:331–340.
13 Kuliński S: Monografia Brześcia Kujawskiego. Włocławek, 1935.

Prof. Elżbieta Żądzińska
Department of Anthropology, University of Łódź
Banacha 12/16
PL–90–237 Łódź (Poland)
Tel. +48 42 635 44 55, Fax +48 42 635 44 13, E-Mail elzbietz@biol.uni.lodz.pl

Koppe T, Meyer G, Alt KW (eds): Comparative Dental Morphology.
Front Oral Biol. Basel, Karger, 2009, vol 13, pp 178–183

Deciduous Tooth Growth in an Ancient Greek Infant Cemetery

Charles FitzGerald[a] · Simon Hillson[b]

[a]Department of Anthropology, McMaster University, Hamilton, Ont., Canada; [b]Institute of Archaeology, University College London, London, UK

Abstract

The Kylindra cemetery on Astypalaia in the Dodecanese, in use 750 BC to 1st century AD, contains a unique skeletal collection. Over 2,400 infant inhumations, each buried in its own clay pot, have been uncovered so far. The skeletal material from each burial is embedded within a ball of accreted earth and since 2001, some 850 infant remains have been recovered and conserved. Most of these died perinatally, but some were very premature babies and some appear to have survived for several months after birth. A study to estimate ages at death of 277 teeth from 107 infants, using microstructural growth markers, is currently underway. One immediate objective is to help resolve the enigma of why such an unusually large number of infants were interred on such a remote Aegean island. Longer term objectives are to reconstruct the sequences of development of the different deciduous tooth types, providing new standards of growth for long bones, the skull and the dentition. This paper presents an interim report on the findings from the histological study, which has analysed 68 teeth from 36 individuals so far. Five of the 36 infants survived birth, three dying within the first 3 weeks of life and the other two surviving for about 3 months. Average appositional growth rates were found to be 3.6 μm/day, and initial mineralisation was found to be well below the figures in Sunderland and coworkers' study in 1987.

In 1999, workers from the 22nd Ephorate of Prehistoric and Classical Antiquities uncovered an extraordinary ancient cemetery on the Greek island of Astypalaia in the Dodecanese. One of the unique features of this cemetery, known as Kylindra, is that it contains only infants, most of them of perinatal age. Another of its impressive attributes is its size: 2,400 burials have been excavated so far, with, in addition, an unknown number extending to both sides and downslope of the area under excavation. However, from a bioanthropological point of view, the most remarkable aspect of Kylindra is that each young child was individually interred in its own amphora or pot. The association of dental and skeletal remains from one infant, instead of the more usual commingling found in ancient sites, offers an unparalleled opportunity to study intra-population variation in development and to make a direct comparison of individual growth trajectories in teeth and bones. Unfortunately, over time, most of these pot urns collapse in the ground and become filled with earth, which then cements together with carbonate to make a solid large ball. Recovering the delicate child's skeleton from this ball is technically challenging and since work began in 2001, only about 850 infant skeletons have so far been extracted from their pots, conserved and stored.

Astypalaia is a small island, having a population today of just over 1,200 permanent residents. Even in Classical times, when it was an independent state flourishing on trade and described in Pliny's Natural Histories [2, 3], the Kylindra cemetery seems much larger than expected from estimates of population size over its period of use, whose dates range from about 750 BC to the first century AD, with most from the period 600 to 400 BC. Still unanswered by archaeologists is the question of why so many infants were interred in such an unusual way on such an isolated island over such a long period of time. The hypotheses explaining Astypalaia's infant cemetery are the subject of controversy, ranging from a sanctuary for mothers with difficult pregnancies to a sacrificial assemblage, but no clear answer has yet emerged.

Approaching a solution to this archaeological enigma requires that precise and accurate ages be assigned to each infant's remains. The bioanthropological objective of studying development also requires the establishment of very accurate chronologies. Unfortunately, macroscopic methods, recording 'stages' of tooth formation or measuring crown heights, suffer problems of observer error and biased dependence on the values of the reference sample and they have not proven useful with the Kylindra material, yielding estimates that are often clearly wrong. The only technique that offers the accuracy required to achieve both of these objectives is odontochronology, the assignment of chronology to dental development events based on interpreting incremental microstructures (see [4, 5] for examples of the accuracy achievable with these techniques). Therefore, it was decided to undertake an odontochronological study of a subsample of Kylindra teeth.

Materials and Methods

Materials

The Kylindra subsample for this histological study consisted of 277 deciduous teeth from 107 infants chosen from among the 850 that have so far been removed from their burial pots. This paper will report on the teeth that have been analysed to date: 68 teeth from 36 individuals.

Deciding what teeth to select was important. The first criterion was to choose teeth only from individuals that had a fairly complete skeleton since it is intended to use the chronologies that are derived in this project in subsequent growth studies. Since the infants in this cemetery are so singular that we wished to maintain alternative teeth in case of damage, the second criterion was to select only teeth that had antimeres present in a dentition.

Methods

Teeth were embedded in epoxy, sectioned and lapped to less than 100 micrometers. Although details on odontochronological techniques and the theory supporting their use can be found in many other publications [e.g. 6–9], most of these refer to permanent teeth [for exceptions, see 4, 10]. Only a broad overview and the main differences applying to deciduous teeth will be provided here.

The principle underlying odontochronology is that certain microstructures in enamel (and dentine) are formed with regular periodicities during growth. These microstructures can be used to assign chronology to dental development events. One important microstructure, the cross-striation, can be seen in figure 1. Cross-striations appear as pairs of light and dark bands in polarized light, and their formation is regulated by the circadian rhythm so that each represents 24 h of enamel growth. Counting the number of them along a prism will therefore yield the time in days taken to form that prism. However, in most cases counting is impractical, and in deciduous teeth, it is rarely possible since cross-striations are more difficult to discern than in permanent teeth. The approach taken is illustrated in figure 1. The *apparent*[1] path of a prism is traced and measured and then groups of cross-striations adjacent to the prism, wherever these are most visible, are counted and measured to arrive at daily average growth rates. These are then prorated across the prism length and its total development

[1] Prisms decussate through the enamel mantel, to a greater degree in prisms directly beneath the cusp than those more cervical; however, even lateral prisms decussate to some extent. The course of a prism path seen on a two dimensional image is therefore more apparently continuous than really so since the prism may actually be weaving into and out of the sectioned plane, which itself is not likely in any event to be perfectly planoparallel. Nevertheless, there is a very large corpus of evidence for the efficacy of odontochronology (refs [5] and [4] to cite but two examples; there too many to list in a paper of this length).

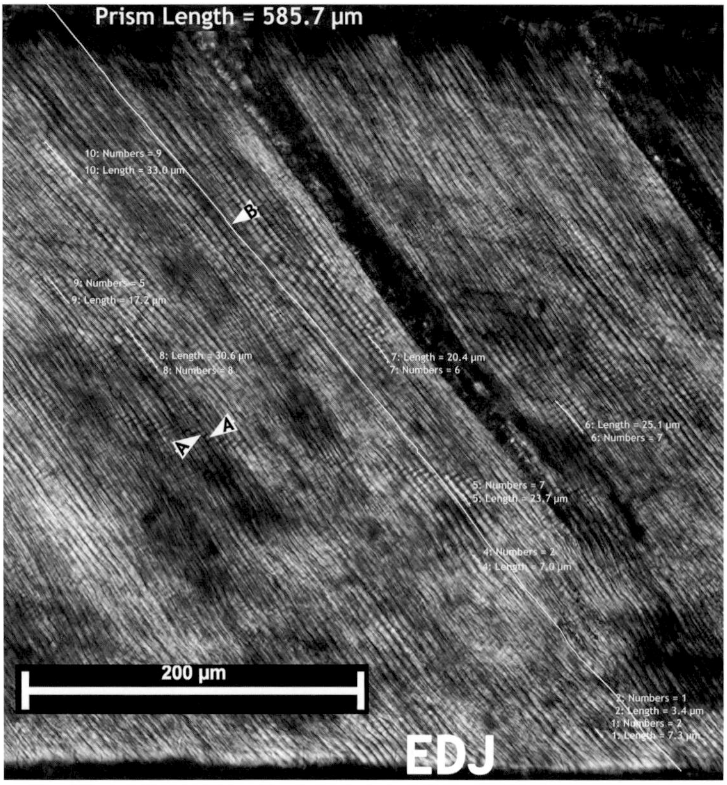

Fig. 1. This is a longitudinal section of prenatal enamel from a deciduous upper second incisor near the cusp tip. The image is taken under polarized light and is demonstrating one approach to odontochronology. Cross-striations, each representing 24 h of enamel growth, can be seen as dark and light bands along prisms, which are running from the enamel dentine junction (EDJ) at the bottom of the image at an approximately 60° angle, to the occlusal surface at the top of the image. The arrows labelled 'A' point to two cross-striations and the one labelled 'B' points to a line that has been traced along the apparent path of one prism. The length of this prism has been annotated by the software used to analyse the image, NIS-Elements BR by Nikon Instruments Inc. On both sides adjacent to the prism can be seen measurements and counts of groups of cross-striations where these are most discernable. These are used to calculate average daily cross-striation rates that are utilized to calculate the time taken for this prism to develop. Note that despite there being no neonatal line in this image, many accentuated striae (also known as Wilson bands) are evident in this prenatal enamel. These are non-regular or pathological striae triggered by some form of external stress.

time is calculated. Another essential microstructure is the neonatal line, which if present, indicates that the infant survived birth, at least for the several days required for the line to form. In odontochronology, the neonatal line 'zeros' chronology and permits estimates of actual time lived. Most other age assessment methods yield biological age, not the true chronological estimates that are possible with these techniques. Essentially, there are two main differences when analysing deciduous rather than permanent teeth: (1) no brown striae of Retzius are found in prenatal enamel, and (2) cross-striations are often more difficult to discern in deciduous teeth.

Results and Discussion

Five of the 36 individuals whose teeth have been analysed survived birth long enough to lay down neonatal lines. Two of these lived for three months

Table 1. Crown formation results for infant with four teeth in the sample

Specimen	Tooth	Crown formation, months		
		prenatal	postnatal	total
F222a54	D R Upp PM3	5.2	0.7	5.9
F222a81	D R Low I1	5.3	0.7	6.0
F222a82	D R Low I2	5.7	0.7	6.4
F222a74	D L Low PM3	5.7	0.7	6.4

Table 2. Average daily growth rates by tooth types (pooled jaws)

Tooth	n	Average growth rate	
		μm/day	SD
D 1st incisor	27	3.7	0.16
D 2nd incisor	28	3.7	0.28
D canine	1	3.9	
D 3rd premolar	10	3.4*	0.29
Overall	66	3.6	0.25

*Significantly different to I1 and I2.

and the other three died within the first month of birth. One of these contributed four teeth, the most from any one individual analysed so far, allowing us to verify the precision of our computed ages at death. Results for this infant are shown in table 1. Two aspects of this result validate the odonto-chronological techniques in this study: (1) neonatal lines were found in all four teeth, and (2) the estimated deaths are within a few days of each other for all of the teeth, at just over three weeks old. As a corollary, it is worth noting all other individuals with more than one tooth in the sample have shown the same consistency (i.e. all teeth from one dentition either showed neonatal lines, or they did not).

So far as we are aware, no histological studies have been published giving detailed calculations of daily appositional growth rates for deciduous teeth comparable to studies for the permanent dentition [8, 11]; however, the 'received wisdom' from the literature is that deciduous teeth grow at a rate of 4 micrometers per day. Table 2 shows our results of daily growth rates by tooth type, with an overall average of 3.6 μm/day for our 66 teeth (note that upper and lower jaw rates for each type showed no significant differences and so they were pooled for this table). Incisor rates of 3.7 μm/day differed significantly from those of third premolars at 3.4 μm/day. It can also be seen in table 2 that standard deviations are in general quite low, but are lower in both upper and lower first Incisors than other tooth types, and lower for mandibular than maxillary teeth. The particularly low standard deviation in first incisors may be the result of two factors related to their simpler geometry. First, this may make the interpretation of microstructures easier and second, since precise mid-cuspal sectioning is very important it may be easier to be more accurate than in teeth with a more complex geometry, like deciduous premolars. Note that, given the overall standard deviation of 0.25 and a large n of 66, this translates into tight confidence intervals around this mean (the 95% CI is 3.58–3.71 μm/day).

Table 3 shows the ages before birth for the initiation of crown formation by tooth type, compared to a study in the literature [1]. In four of the tooth types Kylindra infant teeth began to initiate growth outside of and below the ranges in

Table 3. Comparison of Kylindra vs. Sunderland et al. calcification initiation times

Tooth	n	Initiation of calcification before birth, months		Within range? Differences in means
		Kylindra	Sunderland et al., 1987	
D Upp I1	15	4.6	6.25–5.25 (mean = 5.8)	no, –1.2
D Upp I2	14	5.0	6.00–4.75 (mean = 5.4)	yes, –0.4
D Upp PM3	6	5.3	6.00–5.25 (mean = 5.6)	no, –0.3
D Low I1	12	4.4	6.25–5.25 (mean = 5.8)	no, –1.4
D Low I2	14	4.7	6.00–4.75 (mean = 5.4)	no, –0.7
D Low C	1	5.0	5.25–4.50 (mean = 4.9)	yes, +0.1
D Low PM3	4	4.9	6.00–5.25 (mean = 5.6)	no, –0.7

the Sunderland et al. [1] study and in all but one tooth type, Kylindra initial mineralisation was substantially below the mean ages of the ranges. The only tooth type to exceed the Sunderland and colleagues' mean was the lower canine, which consisted of a single tooth (and even then, it was only 0.1 of a month above).

However, the fact that Kylindra tooth initiation appears to be much earlier than that of the Sunderland et al. [1] infants is not surprising. First, the Sunderland et al. [1] study was based on modern British fetuses, most resulting from abortions carried out for social reasons and of these, any that had gross or microscopic anomalies, chromosomal anomalies, or inadequate histories were excluded from the study. This means that infants with aberrant foetal development were not included by Sunderland et al. [1], but, perforce, might well be in the Kylindra sample since there was no way of identifying them. Secondly, and likely having a greater impact, gestational ages were calculated for each fetus in the Sunderland et al. [1] study based on full obstetric histories accompanied by a pathologist's assessment of the fetus and the placenta, as well as precise crown-rump and other measurements. Ages were expressed as post-menstrual, taken to be two weeks after conception. Fetal jaws were sectioned to determine the degree of mineralisation for each tooth. Therefore, Sunderland et al. [1] were able to confidently express gestation age and locate mineralisation for each tooth against it. The situation with Kylindra was, unavoidably, very different. In Classical Greece, caesarean sections were not carried out, so it must be assumed that every child buried at Kylindra was naturally, vaginally born; but, whether alive or stillborn, or whether premature, or born at or beyond full term cannot, of course, be known. Therefore, in the absence of a neonatal line, we are forced to conclude that death occurred before, during or very shortly after birth, which must also be assumed to have been a normal full term 280 days. This assumption is incorporated in our estimates of initial mineralisation for each tooth, which was calculated by counting backwards from an infant's 'birth' by the crown formation time that had been determined histologically. In table 3 the ages in our study and in Sunderland et al.'s [1] have been adjusted to be on the same base, allowing an unbiased comparison (at least in theory). However, this in no way compensates for the fact that many of the Kylindra infants were premature and did not survive to full term. Since we only have neonatal lines in 5 infants, we can presume that many, if not all of the remaining 31 have been over-aged. In fact, the differences to the Sunderland et al. mean ages of tooth initiation, the last column in table 3, which average 0.82 months, must represent, in part, the number of months that these children were born premature.

Conclusions

This is a preliminary report of a large study. We are still not in a position to unravel the archaeological enigma of Kylindra, although it appears that so far only a small proportion, about 15% (5/36), of the infants who where interred there had survived birth. Although we have collected much data on deciduous tooth development (including some, such as the analysis of Wilson bands, not reported here) we have not yet begun to correlate these with the data from other skeletal elements for these same infants. More meaningful conclusions will have to await further analysis.

Acknowledgments

The authors would like to acknowledge the financial support of the Social Sciences and Humanities Research Council of Canada and the collaboration and assistance of Maria Kollia and the 22nd Ephorate of Prehistoric and Classical Antiquities of Greece. We also wish to pay our respects to the original Principal Investigator of this project, Shelley Saunders, who tragically died during the study.

References

1 Sunderland EP, Smith CJ, Sunderland R: A histological study of the chronology of initial mineralization in the human deciduous dentition. Arch Oral Biol 1987;32:167–174.
2 Morris I: Death-Ritual and Social Structure in Classical Antiquity. Cambridge, Cambridge University Press, 1992.
3 Garland RSJ: The Greek Way of Death. London, Duckworth, 1985.
4 FitzGerald CM, Saunders SR: Test of histological methods of determining chronology of accentuated striae in deciduous teeth. Am J Phys Anthropol 2005;127:277–290.
5 Antoine DM: Evaluating the Periodicity of Incremental Structures in Dental Enamel as a Means of Studying Growth in Children from Past Human Populations. London, University College London, 2000.
6 FitzGerald CM, Rose JC: Reading between the lines: dental development and subadult age assessment using the microstructural growth markers of teeth; in Katzenberg MA, Saunders SR (eds): The Biological Anthropology of the Human Skeleton, ed 2. New York, Wiley, 2008, pp 237–264.
7 Dean MC, Beynon AD: Histological reconstruction of crown formation times and initial root formation times in a modern human child. Am J Phys Anthropol 1991;86:215–228.
8 Reid DJ, Dean MC: Variation in modern human enamel formation times. J Hum Evol 2006;50:329–346.
9 Hillson SW: Dental Anthropology. Cambridge, Cambridge University Press, 1996.
10 FitzGerald CM, Saunders SR, Bondioli L, Macchiarelli R: Health of infants in an Imperial Roman skeletal sample: perspective from dental microstructure. Am J Phys Anthropol 2006;130:179–189.
11 Beynon AD, Dean MC, Reid DJ: On thick and thin enamel in hominoids. Am J Phys Anthropol 1991;86:295–309.

Dr. C. FitzGerald
Department of Anthropology, McMaster University
1280 Main St W
Hamilton, ON Canada L8S 4L9 (Canada)
Tel. +1 905 525 9140, Ext. 26539, Fax +1 905 522 5, E-Mail cfitzg@mcmaster.ca

Koppe T, Meyer G, Alt KW (eds): Comparative Dental Morphology.
Front Oral Biol. Basel, Karger, 2009, vol 13, pp 184–189

Tooth Cementum Annulation Method: Accuracy and Applicability

Zuzana Obertová · Michael Francken

Institut für Rechtsmedizin, Uniklinikum Düsseldorf, Düsseldorf, Germany

Abstract

Tooth cementum annulation (TCA) technique has been a frequently discussed method for the individual age estimation. Conflicting statements on its accuracy and applicability in previous publications have provoked our research. The accuracy and bias of the TCA age estimates were examined in a sample of 116 teeth from 65 individuals of known age and sex from the anatomical collection of the University of Tübingen (Germany). Incremental lines were counted on enhanced digital images of undecalcified, unstained, 60–80 μm thick cross-sections from the middle third of the root of single-rooted teeth. Maximal line counts resulted in age estimates that correlated best with the real age of the specimens. In this sample, this argument is supported by the observation that the mean number of lines increased significantly from the most cervical to the most apical section. Reasonably accurate age estimates based on TCA counts were only obtained in young adults. Both accuracy and bias continuously decreased with the increasing age of the individuals. A considerable underestimation of age occurred in individuals older than 40 years. Due to the conflicting results on the accuracy of the TCA technique this method should be used for age estimation only in association with the macroscopic examination.

Age estimation in human skeletal remains poses a significant problem in historical and forensic anthropology. Macroscopic examination yields reasonably accurate age estimates in juveniles, but the accuracy rapidly deteriorates with increasing age [1]. Several recent studies have recommended the tooth cementum annulation (TCA) method as the technique of choice for accurate age estimation in adults [2–5]. This histological method is based on the examination of hyper- and hypomineralized layers in the acellular extrinsic fiber cementum [2, 6, 7]. These microscopic layers are visible on the cross-sections of tooth roots as alternating light and dark bands, a pair of which is assumed to correspond to 1 year in an individual's life [2, 4, 6].

In this paper the accuracy and bias of the age estimates based on TCA counts are tested in a historical sample of known age and sex.

Material and Methods

Single-rooted permanent teeth were obtained from 65 anatomical specimens of known age and sex from the Osteological Collection of the University of Tübingen (Germany). The sample originates from the turn of the 20th century AD. The age of 48 males and 17 females ranges from 20 to 75 years with a mean of 42.2 years. Fifty skulls come from Europe (mostly Germany) and 15 from Sri Lanka (table 1).

Table 1. Age, sex and ancestry distribution of specimens and teeth.

Age	Europe		Sri Lanka	
	males (n teeth)	females (n teeth)	males (n teeth)	females (n teeth)
20–29 years	7 (14)	4 (5)	0 (0)	6 (12)
30–39 years	6 (13)	1 (2)	5 (9)	2 (3)
40–49 years	7 (10)	1 (2)	2 (4)	0 (0)
50–59 years	13 (26)	1 (2)	0 (0)	0 (0)
60–69 years	6 (9)	2 (3)	0 (0)	0 (0)
70–79 years	2 (2)	0 (0)	0 (0)	0 (0)
Total	41 (74)	9 (14)	7 (13)	8 (15)

One hundred and sixteen maxillary and mandibular incisors, canines and premolars were extracted from intact alveolae with dental forceps. The extracted teeth were free of any visible pathological changes (caries, periapical and periodontal disease, trauma).

Each tooth was completely embedded into a block of epoxy resin (resin Biodur® E12, hardener E1) prior to cutting. After the tooth crown was removed, a series of five cross-sections of 60–80 µm was prepared. Undecalcified and unstained teeth were sectioned in the cervical-apical direction using the Leica 1600 microtome with a diamond-coated saw. The sections were cleaned in distilled water, and mounted serially onto glass slides with Eukitt for the microscopic analysis. Thin sections were examined in transmitted light at 200- to 400-fold magnification, using a Zeiss AxioImager A1 microscope.

Digital images of four quadrants (mesial, distal, buccal/labial, lingual) were taken with the AxioCam Mrc Rev. 3 camera. The images were saved under coded file names to ensure blinded data collection. Before counting the images were enhanced by contrast improvement and adjusted through the grey-scale gradation using Adobe Photoshop software. Starting from the eruption line [7], dark lines were counted at the monitor (17', 1,440 × 900 pixel resolution). In an ideal case, 60 counts per tooth (4 quadrants × 5 sections × 3 counts) were obtained. In cases of good visibility or on the contrary, if parts of the images were blurred, several images per quadrant were taken and included in the counting.

Due to the coding of the image files, any information on the tooth, including tooth type, quadrant, section rank and age was not available to the investigator during the counting. Histological age was assessed by adding tooth- and sex-specific mean eruption age [8] to the counted number of incremental lines.

Statistical analysis was performed by STATISTICA 5.0 and SPSS 15.0 software.

As a measure of accuracy the mean absolute error [MAE = ΣAbs(estimated age – real age)/n] was calculated. It indicates how close the estimate is to the actual value without reference to over- or underprediction. The over- and underestimation is expressed by bias, which is defined as the mean error [ME = Σ(estimated age – real age)/n].

The intra-observer variability of 8 % was established in a random sample of 120 images from 10 individuals. The inter-observer variability between the authors has been tested in previous studies and has not exceeded 10%.

Results

For the purpose of this study, the results are presented regardless of sex and ancestry of the specimens.

In table 2, the values of the mean error and mean absolute error are depicted by age groups along with the paired t test significance levels of the difference between the TCA estimated and the real age. The maximal and mean values are presented here as substitutes for the line counts. Maximal values resulted in more accurate and less biased estimates compared to the mean values. Except for the maximal values in the 20- to 29- and 30- to 39-year-olds there is consistently a

Table 2. Mean error (ME) and mean absolute error (MAE) of maximal and mean values by age group.

Age	n	Maximum		Paired t test p[a]	Mean		Paired t test p[a]
		ME	MAE		ME	MAE	
20–29 years	31	1.1	5.3	0.372	−2.3	4.7	0.018
30–39 years	27	−2.3	7.0	0.166	−7.7	7.8	0.000
40–49 years	16	−15.8	17.8	0.000	−21.0	21.0	0.000
50–59 years	28	−22.0	22.1	0.000	−26.8	26.8	0.000
60+ years	14	−32.0	32.0	0.000	−36.7	36.7	0.000
Total	116	−11.6	14.7	0.000	−16.2	16.9	0.000

[a] Significance level of the difference between the TCA estimated and real age.

statistically significant difference between the estimated and the real age. Both accuracy and bias decreased with advancing age. A considerable underestimation of age occurred in individuals older than 40 years.

Individual differences between the real age and the TCA estimated maximal and mean ages are depicted in figure 1. If the error range of ±2 years off the real age is considered in 20- to 39-year-old individuals' age estimates based on maximal line counts show a significantly higher hit rate (Fisher exact test p = 0.0426) than estimates based on mean line counts. A number of previous studies [3–5] cited this age range as the measure of accuracy of the TCA method. In reality, this age range covers only the variability in individual tooth eruption age.

Statistical analysis (paired-sample t test) of consecutive sections revealed a mean difference of 3.6 lines between the most apical (5th) and the most cervical (1st) section (p = 0.0000); the cervical section showing fewer lines. A similar pattern of mean differences between the number of lines was further observed between the 5th and 2nd sections (2.5 lines; p = 0.0000) and the 5th and 3rd sections (2.2 lines; p = 0.0000). No significant differences occurred between the 5th and 4th section, the 1st and 2nd, and the 1st and 3rd section.

Discussion

Studies on the TCA technique differ significantly according to the statistical parameters applied for the final age estimates as well as to the statements on the accuracy of the results. Nonetheless, one step of the TCA analysis has been quite consistent: the incremental lines are counted on a series of sections, and each section is counted several times. Thus, a statistical parameter is needed that most accurately summarizes all these counts and leads to an age estimate that correlates well with the real age of the individual.

Most frequently, the final TCA age estimates have been derived from the maximal number of lines [9–12] or from the arithmetic mean of all counts [5, 13–15]. Few studies used mode or median [4, 16–17]. In the present study, age estimations based on maximal line counts correlated best with the real age. Mean values resulted in an underestimation of age in every decade.

In this sample, the number of lines increased significantly from the most cervical to the most apical section. This observation supports the argument for the use of the maximal line counts in the TCA age estimation. It also explains why the calculation of the arithmetic mean would result in an incorrect age estimate [see also 12, 16].

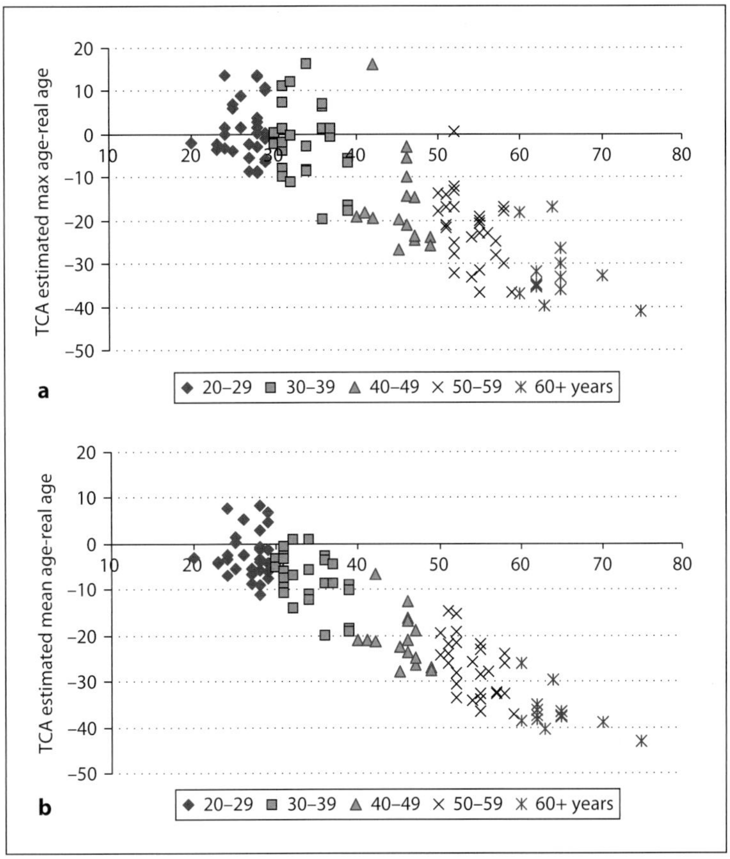

Fig. 1. Individual differences between real age and TCA-estimated maximal age (a) and mean age (b) by age group.

A mean difference of ± 2–3 years between the TCA estimated age and the real age of the individual has been previously suggested [3–5]. As opposed to these studies the results of this paper confirm the conclusions of authors who refer to a moderate [13–14, 18] or even poor correlation between TCA age and real age [9, 16]. In the present sample, reasonably accurate age estimates were only obtained for young adults (20–39 years old).

A continual decrease of accuracy and bias of the TCA age estimates has been observed with advancing age. While some researchers argue that the TCA method leads to accurate estimates regardless of age of the individual [3–5], the majority reports that the difference between the TCA estimated and real age increases with advancing age [9, 10, 13, 15, 18]. Since dental health progressively deteriorates with increasing age of the individual it should not be surprising that the deviations in the TCA counts increase as well [see also 6, 15, 17].

The lack of a perfectly healthy control sample denotes a major drawback in the testing of the TCA method. Although some of the clinically extracted teeth are free of pathological changes the extraction occurs as a therapeutic measure, implying that the dentition has been subject to pathological processes. Similarly, the teeth in the present sample were free of visible pathological conditions but many specimens, particularly those of advanced age, showed signs of dental disease. Nevertheless, the utilization of the TCA method in practice requires that specimens with dental pathologies are included in the studies so their effect on the resulting age estimates can be assessed.

Another substantial disadvantage of this method is the fact that the physiological processes responsible for the formation and deposition of cementum in alternating light and dark bands are essentially unknown. Recently, some basic research has been conducted on the mechanisms of the formation of the different cementum types, and on identification of proteins specific to cementum [19–22]. The clarification of the role and effect of these biological variables will consequently improve the conditions for tests of the applicability of the TCA method in individual age estimation. Until then, due to the conflicting results on the accuracy of the TCA technique researchers should remain critical, and if possible use this method for age estimation only in association with the macroscopic examination.

Conclusions

In the present study, the accuracy of the TCA age estimates was tested in a sample of 116 teeth from 65 individuals of known age and sex from the anatomical collection (19th/20th centuries AD) of the University of Tübingen.

Maximal line counts result in age estimates that correlate best with the real age of the specimens. In this sample, this argument is supported by the observation that the number of lines increased significantly from the most cervical to the most apical sections. Reasonably accurate TCA age estimates were only obtained in young adults. Both accuracy and bias continuously decreased with increasing age of the individuals. A considerable underestimation of age occurred in individuals older than 40 years.

Further testing of the applicability of the TCA technique for individual age estimation will be necessary; particularly after the biological mechanisms responsible for the deposition of the cementum are better understood. Until then, the TCA method should be used for age estimation only in association with the macroscopic examination.

Acknowledgements

This study was supported by the project grant for young researchers from the University of Tübingen.

We would like to thank Prof. J. Wahl, who generously allowed access to the laboratory equipment at the Landesdenkmalamt Konstanz. We are grateful to the reviewers for their helpful comments on an earlier version of the manuscript.

References

1 Buikstra JE, Ubelaker DH (eds): Standards for Data Collection from Human Skeletal Remains. Fayetteville, Arkansas Archaeological Survey Research Series 44, 1994.
2 Stott GG, Sis RF, Levy BM: Cemental annulations as an age criterion in forensic dentistry. J Dent Res 1982;61:814–817.
3 Grosskopf B: Individualaltersbestimmung mit Hilfe von Zuwachsringen im Zement bodengelagerter menschlicher Zähne. Z Rechtsmed 1990;103:351–359.
4 Kagerer P, Grupe G: Age-at-death diagnosis and determination of life-history parameters by incremental lines in human dental cementum as an identification aid. Forensic Sci Int 2001;118:75–82.
5 Wittwer-Backofen U, Gampe J, Vaupel JW: Tooth cementum annulation for age estimation: results from a large known-age validation study. Am J Phys Anthropol 2004;123:119–129.
6 Klevezal GA, Kleinenberg SE: Age Determination of Mammals from Annual Layers in Teeth and Bones. Jerusalem, IPST Press, 1969.

7 Kagerer P, Grupe G: On the validity of individual age-at-death diagnosis by incremental lines counts in human dental cementum: technical considerations. Anthropol Anz 2001;59:331–342.

8 Schumacher GH, Schmidt H, Böring H, Richter W: Anatomie und Biochemie der Zähne. Stuttgart, Gustav Fischer, 1990.

9 Lipsinic FE, Paunovich E, Houston GD, Robison F: Correlation of age and incremental lines in the cementum of human teeth. J Forensic Sci 1986;31:982–989.

10 Kvaal SI, Solheim T: Incremental lines in human dental cementum in relation to age. Eur J Oral Sci 1995;103:225–230.

11 Wittwer-Backofen U, Buba H: Age estimation by tooth cementum annulation: perspectives of a new validation study; in Hoppa RD, Vaupel JW (eds): Paleodemography: Age Distributions from Skeletal Samples. Cambridge, Cambridge University Press, 2002, pp 107–128.

12 Lippitsch A, Grupe G: Variability of the apposition of the acellular, extrinsic fibre cementum and its influence on the tooth cementum annulation technique in humans: the influence of physical demands and functional morphology. Doc Archaeobiol 2007;5:87–112.

13 Condon K, Charles DK, Cheverud JM, Buikstra JE: Cementum annulation and age determination in *Homo sapiens*. II. Estimates and accuracy. Am J Phys Anthropol 1986;71:321–330.

14 Jankauskas R, Barakauskas S, Bojarun R: Incremental lines of dental cementum in biological age estimation. Homo 2001;52:59–71.

15 Pilloud S: Lässt sich mittels der Altersbestimmung anhand des Zahnzementes auch bei älteren Individuen ein signifikanter Zusammenhang zwischen histologischem und reellem Alter finden? Anthropol Anz 2004;62:231–239.

16 Renz H, Radlanski RJ: Incremental lines in root cementum of human teeth: a reliable age marker? Homo 2006;57:29–50.

17 Francken M, Wahl J: Die Zahnzementannulation im Vergleich zu konventionellen Methoden der Sterbealtersbestimmung an den bandkeramischen Skelettresten aus dem Gräberfeld von Schwetzingen. Fundberichte Baden-Württemberg 2007;29:747–761.

18 Meinl A, Huber CD, Tangl S, Gruber GM, Teschler-Nicola M, Watzek G: Comparison of the validity of three dental methods for the estimation of age at death. Forensic Sci Int 2008;178:96–105.

19 Bosshardt DD, Schroeder HE: Initiation of acellular extrinsic fiber cementum on human teeth: a light- and electron-microscopic study. Cell Tissue Res 1991;263:311–324.

20 Bosshardt DD, Selvig KA: Dental cementum: the dynamic tissue covering of the root. Periodontology 2000 1997;13:41–75.

21 van den Bos T, Beertsen W: Alkaline phosphatase activity in human periodontal ligament: age effect and relation to cementum growth rate. J Periodont Res 1999;34:1–6.

22 Saygin NE, Giannobile WV, Somerman MJ: Molecular and cell biology of cementum. Periodontology 2000 2000;24:73–98.

Dr. rer. nat. Zuzana Obertová, PhD
Institut für Rechtsmedizin, Uniklinikum Düsseldorf
Moorenstrasse 5
DE–40219 Düsseldorf (Germany)
Tel. +49 21181 18294, Fax +49 21181 19366, E-Mail zuzana.obertova@med.uni-duesseldorf.de

Koppe T, Meyer G, Alt KW (eds): Comparative Dental Morphology.
Front Oral Biol. Basel, Karger, 2009, vol 13, pp 190–196

Nothing in Nature Is as Consistent as Change

Kurt W. Alt · Anne Rossbach

Institute of Anthropology, Mainz University, Mainz, Germany

Abstract

Dentition, as a mechanically stressed part of the orofacial system, is subject to physiological wear processes that affect the occlusal surface, the cutting-edge and the approximate contact points of teeth. The reasons are abrasive food particles, tooth contacts during chewing as well as erosion. Up until the Middle ages and even further on, both the deciduous and the permanent dentition were, depending on age, subject to distinct hard tissue defects. These regularly led from normal overbite, which develops during dentition, to a pronounced edge-to-edge bite. In dentistry this known phenomenon is widely interpreted as a pathological adaptation. Due to specific subsistence conditions and dietary habits in food intake and preparation abrasive changes can be found in the dentition of our ancestors, beginning with the history of humanity up until historic times. However, hardly in today's population. Abrasive food particles and erosion are the main factors that cause wear in dental enamel. We analyzed occlusal hard tissue changes that led to edge-to-edge-bite in chronologically scattered skeletal series from different regions in Germany. The sample consists of both males and females from varying age groups. The skulls were photographed in standardized positions and radiographically examined. The results show that dental wear is a natural, age-dependent process which does not lead to pathological changes. Crowding and contact surface caries can even widely be impeded through abrasion. Therefore dental wear is a natural process that has only been prevented by 'civilization' in the past two centuries. Edge-to-edge-bite is still the preferable occlusion in man.

Dental Remains as an Object of Studies about Human Evolution and Life History

The dentition is one of the most informative parts of excavated skeletons found in (pre)historic sites. Why is this the case? Due to their mineral content, teeth are the hardest structures in the human body. They are more durable than skeletal remains, they show less post-mortem taphonomic processes, and they can be directly observed and evaluated in both living and past populations. Due to the good preservation and the large amount of information they retain, teeth can provide a bio-historic source particularly well suited to foster insight into important aspects of human life history [1–3] such as growth and development, individuality, health and pathology, the relationships between nutrition, behaviour and the dentition, as well as geographic and genetic variation [4]. From the attained findings it is possible to draw conclusions that may be relevant for contemporary clinical dentistry.

In (pre)historic skeletal material, dating up until the Middle Ages and even further on, both the deciduous and the permanent dentition were subject to distinct hard tissue defects. These regularly led from normal overbite, which develops during dentition, to a pronounced edge-to-edge bite. Changes like these follow natural processes that do not lead to pathological alterations but are a sign of functional adaptation.

Tooth Wear in the Past and Present

Causes for tooth wear are food, grit (abrasion), opposing tooth surface contacts (attrition), chemical processes not involving bacterial action (erosion), and task-related wear (i.e. teeth as tool) [5, 6]. Dental attrition is caused by tooth-tooth contact of neighbouring and opposing teeth. As a result, wear facets form on the occlusal zones and the interstitial contact points of the teeth, respectively. Dental abrasion describes the loss of dental hard tissue by abrasive particles, e.g. in foods, that affect the teeth during chewing. Nowadays, abrasion is mainly caused by grinding particles in toothpaste [2, 5, 7]. Erosive processes that are due to acidic foods (e.g. lemons), sour drinks (e.g. wine, soft drinks) or vomit (bulimia) also cause loss of dental hard tissues. However, in archaeological context it is difficult to differentiate between the single causes that lead to the loss of dental hard tissue.

When observing teeth from all ages of human history, starting with the early *Homo erectus* finds to the beginning of the food-producing economy (early farmers), the antiquity and the middle ages up until modern times, it can be asserted that dental wear is a universal occurrence [2, 8–11]. However, tooth wear is not restricted to humans but is also found as a natural process on the teeth of mammals. In archaeozoology and palaeontology the degree of dental wear is used to determinate the age of certain species [7, 12, 13].

With the introduction of the neolithisation and the associated change in food preparation the pattern of dental wear also changed, but neither the basic process nor the degree of wear [14, 15]. Not until the beginning of the industrialisation, which besides many other consequences also greatly affected food processing, the degree of dental wear decreased. Today, it is absent except for parafunctions, e.g. bruxism. Therefore, excessive wear on teeth is a rather rare finding in contemporary dentistry.

Natural Tooth Wear in Naturally Living Indigenous Populations

The close the relationship between tooth wear and lifestyle is documented by studies among indigenous populations who live largely unaffected of today's civilization and therefore are very well suited for comparison with (pre)historic times. Pereira and Evans [16] as well as Garve and von Puttkamer [17] found impressive examples of natural tooth wear patterns during their examination of Yanomami, in Papua New Guinea and with pygmy. The Yanomami are an Indian population from northern Brazil, whose lifestyle is aboriginal and free of cultural adjustment. Within this indigenous group all individuals show a high degree of dental wear. Neither geographically nor compared to prehistoric and historic populations differences worth mentioning concerning form and degree of wear can be found. This results in the fact that with reaching mature age, the occlusal cusps of the teeth are almost worn down. In addition, due to a concurrent forward shift of the lower jaw, a frontal edge-to-edge bite occurs. With increasing age the occlusal surface of the teeth are completely levelled and edge-to-edge bite is the natural consequence of this process. The large degree of dental wear is due to the high consumption of raw vegetables (e.g. roots, corm) and therefore the tougher foods as well as the additionally with abrasives soiled (e.g. sand, dirt) foods.

Is Edge-to-Edge Bite a General Phenomenon?

Tooth wear significantly alters the shape of a tooth crown during its entire period of use. In the 1950s, P. Raymond Begg, an orthodontist, published his studies on occlusion and dental wear in Australian aborigines. Based on the results of his research, he proposed the hypothesis that edge-to-edge bite does not represent a pathological bite form, but would be the correct occlusion for humans, even today [7, 18]. In his opinion, the habitual occlusion was not a static condition, but a continually changing, functional process. The constant loss of both occlusal and interstitial dental hard tissue would therefore be natural.

For the development of an edge-to-edge bite certain positional changes have to occur within the dentition. Two factors are of crucial relevance: (1) a tooth movement that takes place as vertical eruption, and, more importantly (2) a mesial drift of the lower posterior teeth [9, 18]. The first factor is also known as continual growth. It compensates for the height loss of the tooth crown due to wear, thus maintaining the contact between antagonistic teeth. The second factor is known as interstitial tooth wear: The interproximal loss of enamel reduces the width of the teeth. The resulting gaps are closed as the lower teeth move into mesial direction (mesial drift).

The result of the continuous adaptation of the teeth to these morphological changes is an edge-to-edge bite. This kind of occlusion prevents crowding of the front teeth. At the same time sufficient space for a smooth eruption of the third molars – if present – is created. Due to the interproximal tooth loss the mesio-distal length of the dental arch is greatly reduced. Furthermore, the occlusal wear levels the cusps of the posterior teeth; thereby reducing and finally eliminating the fissures as predilection sites for caries. Begg also states that at the same time, the flat occlusal surface of the premolars and molars enhances the masticatory force of the dentition [18].

Occlusal Relationships and Tooth Wear – Clinical Evidence

The term 'dental occlusion' describes any static or dynamic contact between the upper and lower teeth while the jaws are closed. The multitude of theoretically possible occlusions expresses the large range of variation within the dentition [19]. The individual potential of adaptation to ever changing bite conditions (e.g. with age) is so distinct that only with parafunctional dysfunctions is a dental intervention necessary. 'Confusion in occlusion' is the term given by Türp et al. [20] for the discourse of the conceptual theory which concerned dentistry for more than 100 years. In Germany and other countries 'gnathological' orientated concepts of occlusion resulted from this, which conceived the 'anatomical correct occlusion of the permanent dentition' as a static (ideal or harmonic) alignment of the teeth, which changes only slightly during life. The leading representatives of this were B.B. McCollum (1883–1969) and his colleague C.E. Stuart (1890–1982). The latter stated: 'A man without cusps on his teeth is like a man without feet on his legs' [7, p. 312]. This was without a doubt an exaggerated statement but it illustrated the spirit of the time and has survived in parts of dentistry until today.

For the different concepts of occlusion in dentistry, dental anthropology has sound objections, but also valid answers. The attained data and findings are based on decade-long research in the fields of dental evolution and the biologic reconstruction of the lifestyle and living conditions of our ancestors. 'Ideal' and 'harmonic' occlusion represent theoretical models that do not exist in natura. Occlusion is not a stable condition. Like all biological processes, the dentition is subject to constant changes. Continuous wear of teeth is normal in nature. Comparable to degenerative changes in other organs, the deterioration of the dentition, which is linked to age, alters the cutting edges and occlusal surfaces of the teeth and as a consequence the occlusal relationship. This is a sign of functional adaptation. Today this view is favoured in parts of dentistry; consequences for practice, teaching and research have so far not been derived. 'There is no point (...) in wanting to grind a flat

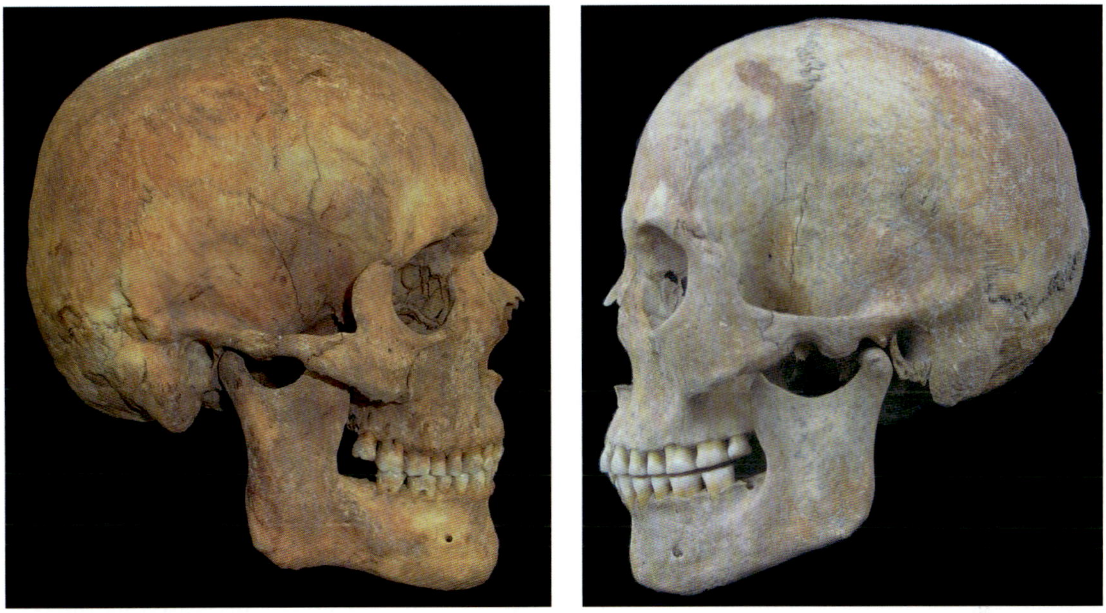

a b

Fig.1. **a** Hettstedt (~1600 AD), individual 289, male 35–40 years. Occlusal cusps not completely worn, edge-to-edge bite. **b** Halberstadt (~5000–4500 BC), Individual 12, male 40–45 years. Occlusal surface completely worn down, edge-to-edge bite and flat occlusal surfaces of the posterior teeth.

teeth articulation in today's dentition, but it is just as wrong, when doing a restoration for middle-aged patients, that already have distinct dental wear facets, to force juvenile, high cusps on them and thereby preventing any kind of self-help of the dentition' [7, p. 316].

In contrast to prehistoric and historic times, almost no wear occurs in teeth of individuals consuming a Western diet, so that the juvenile occlusion is maintained until old age, if no changes caused by parafunctions (e.g. attrition), extractions or age-dependent involutions had occurred [21]. A mesial drift is therefore not possible, since the teeth stay locked in their cusp-fossa position. The consequences of this type of occlusion are manifold. Besides pit caries, crowding of the anterior teeth and a shortage of space for the eruption of the third molars, often temporomandibular joint problems occur [18, 22]. Possibly due to the absence of dental wear the functional structure of teeth, dental periosteum, masticatory muscles and the temporomandibular joints are disturbed in such a way that it results in an increased need for dental treatment. The missing approximal wear can mainly be simulated by clinical interstitial 'slicing' in order to correct crowding of the front teeth. Balancing extractions as postulated in the past are unacceptable to modern dentistry [23, 24].

Comparative Study on the Incidence and Chronological Development of Tooth Wear in Skeletal Remains from Different (Pre)Historic Periods

In an ongoing study, the occlusal relationships in well-preserved skulls from different (pre)historic eras in Central Europe are examined at the Department for Anthropology of the University of Mainz (fig. 1). Both male and female individuals of all age groups including infants and juveniles date between 5000 BC and 1600 AD. Standardized digital photos (lateral view of the skull in the Frankfurt plane, pictures of the occlusal surfaces of the upper and lower jaws) and digital volume tomographies (DVT) of all individuals were taken, as well as measurements of the width of the teeth in mesio-distal and bucco-lingual directions. The dental wear was recorded according to Miles [25], Molnar [26], and Scott [27].

As a preliminary result of the study it seems that the wearing down of teeth is independent of the time period and the sex. Very similar wear facets are found from the Palaeolithic up until early modern times. The edge-to-edge bite occurs depending on the age of the individuals. In principle, the loss of dental hard tissue begins when the teeth reach the occlusal plane. It increases gradually with age, so that by reaching mature age the edge-to-edge bite occurs, which goes along with a complete levelling of the occlusal and incisal surfaces. Normally, the continuous flattening of the cusps causes an increasing group function. Not in a single one of the examined cases does canine guidance seem to have played a role. Infants and juveniles mainly show a normal bite with a vertical overbite as described by Pereira and Evans [16] for 93% of the young Yanomami (fig. 1). According to this the diagnosed edge-to-edge bite in the majority of the adults is not based on a beforehand existing mesial bite (Angle class III) but develops as a result of the continuous dental wear from a normal bite. Therefore, the edge-to-edge bite is clearly not a pathological appearance, as especially gnathology is trying to make credible [28, 29].

Has the Time for a Change of Paradigm Come?

The hitherto existing studies from chronologically and geographically diverse populations consistently show that the loss of dental hard tissue is a natural process whose result is the perpetuation of a fully functional masticatory system during its entire time of use (fig. 2). The adaptations in the dentition therefore do not lead to pathological changes in the position of teeth and jaws [2, 5, 11, 15, 18]. From the view of dental anthropology, dental wear is a 'normal' occurrence. If during the course of use of the teeth occlusal disturbances such as cross-bite or parafunctions do not occur, then mostly by late mature age (35–45 years) the edge-to-edge bite develops from a normal bite and central occlusion. For developing the edge-to-edge bite (normal occlusion), it is a necessary precondition that interproximal attrition with a subsequent mesial drift of the teeth occurs. As a compensation for the occlusal wear, a continuous eruption of the teeth takes place during a lifetime. Dental crowding, shortage of space as well as the risk of fissure caries are greatly decreased by the flattening of the occlusal surface. The masticatory force of the dentition is greatly increased [7, 10, 18, 22]. The loss of dental hard tissue is therefore a physiological process and a sign of functional adaptation.

The anthropological findings on 'occlusal history' of humans could be the start of a dialogue between dental anthropology and dentistry. The goal should be to transfer the research finding to modern dentistry and to develop new impulses and concepts for the dental/orthodontic diagnosis and therapy. This is only possible in an interdisciplinary cooperation.

In summary, we can state that tooth wear is a natural process that does not lead to pathological changes. Generalised tooth wear develops from centric occlusion and edge-to-edge bite is age- but not sex-dependent in an individual. The early frontal overbite changes into edge-to-edge bite during adulthood, and interproximal attrition supports this process by mesial drift of the lateral teeth. Tooth wear is therefore physiological and a sign of functional adaptation. According to Schray [7, p. 312] '(...) the natural tooth wear seems to be the only depletion in the human body that bring advantages'.

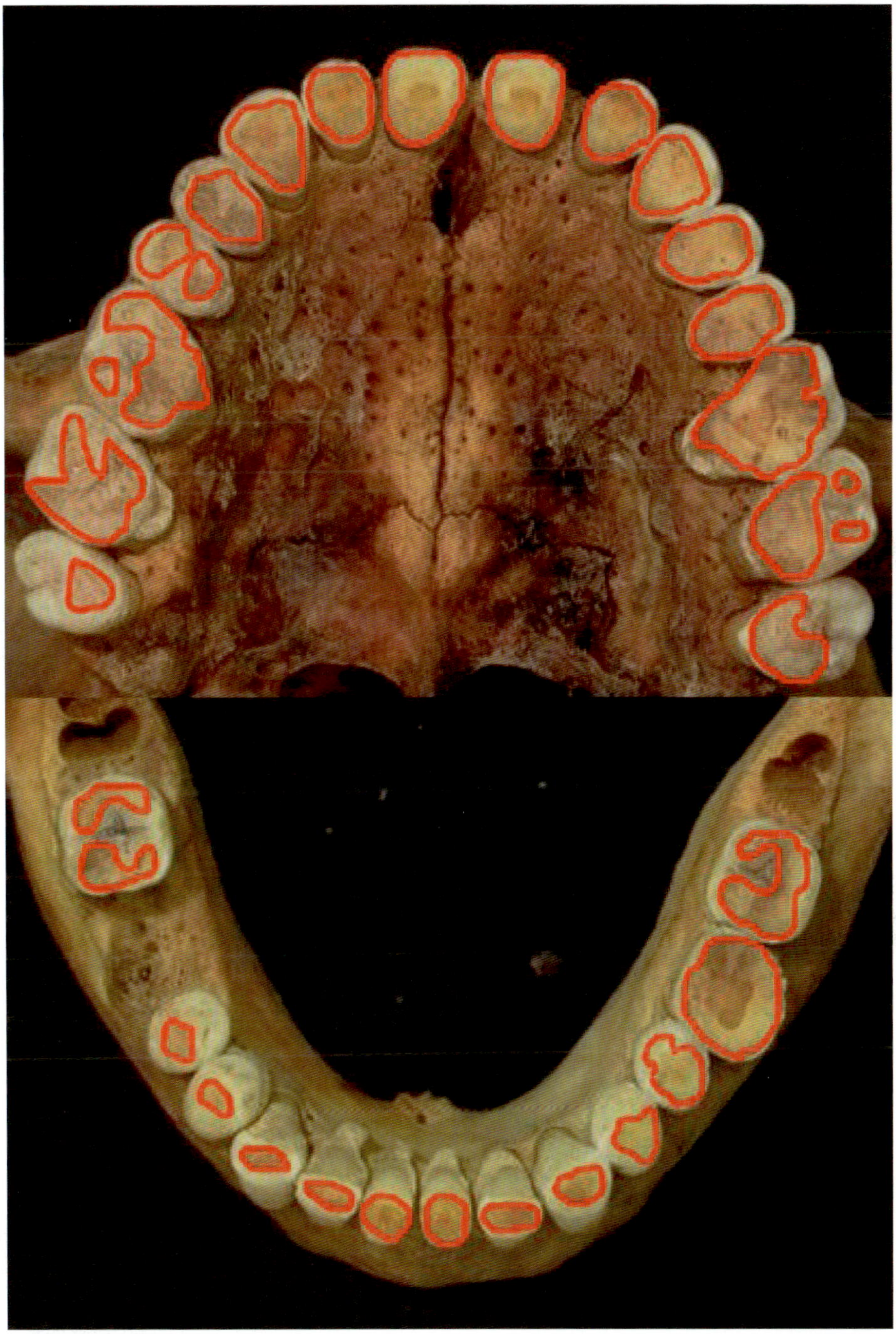

Fig. 2. Hettstedt (~1600 AD), individual 289, male 35–40 years. Tooth 38, 48 extracted for aDNA analysis. Occlusal cusps of the lateral teeth not yet worn flat completely, borders of the wear marked red.

References

1 Alt KW: Die historische Entwicklung der Odontologie und Dental-Anthropologie; in Alt KW, Türp JC (eds): Die Evolution der Zähne. Berlin, Quintessenz Verlag, 1997, pp 47–76.
2 Hillson S: Dental Anthroplogy. Cambridge, Cambridge University Press, 1996.
3 Scott GR, Turner CG: The Anthropology of Modern Human Teeth. Cambridge, Cambridge University Press, 1997.
4 Young WG: Anthropology, tooth wear and occlusion ab origine. J Dent Res 1998;77:1860–1863.
5 Lucas PW: Dental Functional Morphology. Cambridge, Cambridge University Press, 2004.
6 Alt KW, Pichler SL: Artificial modifications of human teeth; in Alt KW, Rösing FW, Teschler-Nicola M (eds): Dental Anthropology. Fundamentals, Limits, and Prospects. Wien, Springer, 1998, pp 387–415.
7 Schray K: Die natürliche Abrasion und ihre gnathologischen Konsequenzen. ZWR 1984;93:310–316.
8 Alt KW, Kaulich B, Reisch L, Vogel H, Rosendahl W: The Neanderthalian molar from Hunas. Homo 2006;57:187–200.
9 Hylander WL: Morphological changes in human teeth and jaws in a high-attrition environment; in Graber TM, Dahlberg AA (eds): Orofacial Growth and Development. Berlin, Mouton de Gruyter, 1977, pp 302–329.
10 Neiburger EJ: Flat-plane occlusion in the development of man. J Prosthet Dent 1977;38:459–469.
11 Neiburger EJ: The evolution of Human occlusion: ancient clinical tips for modern dentistry. Gen Dent 2002;50:44–49.
12 Baker G, Jones LHP, Wardrop ID: Cause of wear in sheep teeth. Nature 1959;184:1583–1584.
13 Pichler S: Zähne und Kiefer in der Archäozoologie- Relevanz und Methoden; in Alt KW, Türp JC (eds): Die Evolution der Zähne. Berlin, Quintessenz Verlag, 1997, pp 731–749.
14 Kaifu Y: Was extensive tooth wear normal in our ancestors? A preliminary examination in the genus Homo. Anthropol Sci 2000;108:371–385.
15 Deter CA: Gradients of occlusal wear in hunter-gatherers and agriculturalists. Am J Phys Anthropol 2009;138:247–254.
16 Pereira CB, Evans H: Occlusion and attrition of the primitive Yanomani Indians of Brazil. Dental Clinics of North America, Symp Alterable Centric Relation in Dentistry, 1975, vol 19, p 3.
17 Garve R, von Puttkamer WJ: Indianer am Amazonas. Adliswil, Tanner Verlag, 1985.
18 Begg PR: Stone age man's dentition. Am J Orthodont 1954;40:4–7.
19 Türp JC, Alt KW, Schindler HJ: Okklusion; in Hugger A, Türp JC, Kerschbaum T (eds): Curriculum Orale Physiologie. Berlin, Quintessenz Verlag, 2006, pp 1–34.
20 Türp JC, Greene CS, Strub JR: Dental occlusion: a critical reflection on past, present and future concepts. J Oral Rehab 2008;35:446–453.
21 Koeck B: Die Abrasion der Kaufläche ein Zeichen funktioneller Anpassung. ZWR 1981;90:50–61.
22 Varrela J: Masticatory function and malocclusion a clinical perspective. Semin Orthodont 2006;12:102–109.
23 Chudasma D, Sheridan JJ: Guidelines for contemporary air-rotor stripping. J Clin Orthodontol 2007;41:315–320.
24 Rosa M, Cozzani M, Cozzani G: Sequential slicing of lower deciduous teeth to resolve incisor crowding. J Clin Orthodont 1994;28:596–599.
25 Miles AEW: The dentition in the assessment of individual age in skeletal material; in Brothwell DR (ed): Dental Anthropology. London, Pergamon Press, 1963, pp 191–209.
26 Molnar S: Tooth wear and culture: a survey of tooth functions among some prehistoric populations. Curr Anthropol 1972;13:511–516.
27 Scott EC: Dental wear scoring technique. Am J Phys Anthropol 1979;51:231–218.
28 Motsch A Funktionsorientierte Einschleiftechnik für das natürliche Gebiss. München, Hanser Verlag, 1977.
29 Bumann A, Lotzmann U: Funktionsdiagnostik und Therapieprinzipien; in Rateitschak KH, Wolf HF (eds): Farbatlanten der Zahnmedizin. Stuttgart, Thieme, 2000, vol 12.

Prof. Dr. Kurt W. Alt
Institute of Anthropology, Mainz University
Colonel-Kleinmann Weg 2
DE–55099 Mainz (Germany)
Tel. +49 06131 39 222 42, Fax +49 06131 39 25 132, E-Mail altkw@uni-mainz.de

Author Index

Subject Index